MAN CHANGES THE WORLD

The Earth, Its Wonders, Its Secrets

Man Changes the World

Published by

The Reader's Digest Association Limited

London New York Montreal Sydney Cape Town

MAN CHANGES THE WORLD
Edited and designed by Toucan Books Limited
with Bradbury and Williams
Written by John Man
Edited by Helen Douglas-Cooper
Picture Research by Marian Pullen

FOR THE READER'S DIGEST UK
Series Editor: Christine Noble
Editorial Assistant: Chloe Garrow
Editorial Director: Cortina Butler
Art Director: Nick Clark

READER'S DIGEST, US
Senior Editor: Fred Dubose
Senior Designer: Judith Carmel
Group Editorial Director, Nature: Wayne Kalyn
Vice President, Editor-in-Chief: Christopher Cavanaugh
Art Director: Joan Mazzeo

Separations: David Bruce Graphics Limited, London

Printed in the United States of America, 1999

Library of Congress Cataloging in Publication Data

Man changes the world.
     p. cm. - -   (The earth, its wonders, its secrets)
   ISBN 0-7621-0115-6
   1. Nature- - Effect of human beings on. 2. Human ecology.
I. Reader's Digest Association. II. Series.
   GF75.M343 1999
     304.2'8 – dc21
                              99-22012

FRONT COVER   *Terraced rice fields cover the side of a hill in Bali. The
Thames barrier protects London from the danger of flooding (inset).*

PAGE 3   *A model of an African elephant, made from old car bumpers,
stands in the Lincoln Park Zoo, in Chicago.*

CONTENTS

THE EARTH TRANSFORMED

In the last 100 years, humans have reached even the remotest places on Earth. Where we have spread, the world has been tamed and transformed. The change has been planet-wide: remarkable in its speed, and almost incredible in its sweep.

Think of the wildest and most remote place on Earth: a place that has never seen a human footprint or heard a human voice, and that has never felt the transforming touch of a human hand; somewhere that has never been explored or examined, used or exploited – in short, an absolute wilderness, where the forces of nature reign supreme.

Where would your ultimate wilderness be? The top of an inaccessible mountain, or the centre of a great desert? A small island in the vastness of the ocean, ringed by precipitous cliffs or by ramparts of living coral?

Or perhaps some point on one of the Earth's great ice caps, more than a mile above the rock beneath, and completely devoid of all signs of life?

Extraordinary though it may sound, none of these places is truly free from human influence. Spurred on by the need for living space, the thirst for knowledge and the sheer spirit of adventure, people have travelled to places that

once seemed forever beyond our reach. And where humans venture, change is rarely far behind.

THE ULTIMATE FRONTIERS

From the air, it seems like a world of infinite emptiness. A dazzling horizon separates the sky from the glistening ice beneath, and the low sun casts razor-sharp shadows from a range of granite peaks that are steep enough to shrug off the year-round snow. The ice itself is almost featureless, and there is nothing to give any idea of scale. The mountains could be 500 ft (150 m) or 5000 ft (1500 m) high, and the extent of the ice field in which they are set is impossible to guess. But as the plane continues its slow descent, something comes into view that reveals the true vastness of the scene: a tiny orange speck set in the immense white void. It is a tent – a temporary summer camp in the Transantarctic Mountains, used by geologists searching the ice for meteorites that have fallen from space.

Few places on Earth are so remote, and few so carefully shielded from human contact. At the end of their brief summer residence, when the constant polar daylight begins to wane and the first sunset approaches, the geologists will pack up their camp, together with the specimens they have collected, and they and their equipment will be flown back to a permanent base. Following the provisions of the Antarctic Treaty – signed in 1959 by 12 nations with a scientific interest in the continent – they will be careful to leave no signs of their sojourn on the ice. That

LOST WORLDS California's Yosemite Valley forms one of the world's most dramatic landscapes. Once an uninhabited wilderness, it now attracts thousands of visitors every year.

FARTHEST REACH
The mountains of eastern Greenland are among Earth's most inaccessible places, yet even here human exploration is several centuries old.

is, except one: a radio beacon that will guide them back the following year.

Some 5000 miles (8000 km) to the north, deep beneath the island of Borneo, a group of people intrudes into very different surroundings. After entering the mouth of a giant limestone cavern, and pausing to admire the aerobatics of tiny swiftlets that raise their young in nests glued to the solid rock, a party of speleologists – people who specialise in the study of caves – adjust their lamps and set off on a journey underground. Within a few minutes the light from the cave's entrance is fading, and the mosses and algae that depend on light for survival have disappeared. The cavers press on through the giant chamber, passing underneath a vast colony of roosting bats that have littered

the cavern with centuries' worth of droppings. Sightless cave crickets, equipped with outsize antennae much longer than their bodies, scuttle away into crevices as they sense the vibrations from the cavers' booted feet, and the bats overhead jostle nervously as the yellow beams of lamplight slice up through the darkness.

Several hours later the cavers are far beyond the realm of these animal inhabitants. The cavern has been replaced by a series of chambers of varying height, and in places the sculpted limestone ceiling is low enough to touch. The party comes to a halt at the lip of a vertical shaft, and then after a pause, two of the team disappear into the darkness down cable ladders that have been unrolled into the void below. There is an anxious wait lasting several minutes, but finally comes a shout: the pioneers have successfully reached the bottom. And here, in one of the most inaccessible places on Earth, their feet form prints in a layer of gravelly sediment, creating tangible proof that humans have arrived.

Far away from this subterranean world, yet another scene unfolds. A research ship pitches and rolls in the grey waters above the continental shelf of eastern North America. Several hundred feet beneath it a two-man submersible, with ocean scientists on board, is closing in on the seabed. At

EXPLORING THE DEPTHS In the comfort of this bubble-shaped observatory, oceanographers can investigate a world that was once beyond the reach of human observers.

*EXPANDING INFLUENCE
A patchwork of fields clothes
the fertile slopes of Costa
Rica's Irazu volcano. Just
50 years ago, most of this area
was uninterrupted forest.*

this depth it is almost dark, and powerful spotlights are needed to illuminate the seafloor. Few divers would care to explore these inhospitable waters, where the temperature is only a few degrees above freezing, and few craft have ever ventured into the turbid gloom. Yet even here, far from the coast, there is already evidence of human activity: crisscrossing the seafloor are parallel gouges scraped out by the nets of trawlers plying the surface far above.

READING THE PAST

These three scenes show how far the human species has gone in its colonisation of the planet Earth. This remarkable process started, from quite insignificant beginnings, some 200 000 years ago in one continent – Africa. During a seemingly inexorable movement across the world, humans have altered their surroundings, and the steady build-up of these changes has created the planet we know today. Many of these changes have been deliberate, carried out with the aim of making life easier. Other kinds of change – from pollution to the introduction of pests and weeds – have been unplanned or accidental, creating problems for us, and also for other forms of life.

In a world as complex as ours, unravelling the story of these changes can be a challenging task, particularly when they took place long ago. Nobody really knows, to take one hotly debated example, how far humans were involved in the disappearance of large prehistoric mammals from North America about 12 000 years ago. Were there enough people to have such a major impact? Was their technology and social organisation up to the task of hunting animals many times their own size? Could these changes have occurred without human intervention at all – perhaps through shifts in climate? Such questions make interpreting the distant past a fascinating but uncertain business.

Other kinds of change, although more recent, are sometimes so well disguised that it takes a trained eye to spot them. To most people, a patch of ground disturbed long ago may look like a natural feature of the landscape. To an archaeologist, its irregular surface may tell a very different story. A field of crops may look smooth and uniform from the ground, but when it is seen from the air slight differences in its growth rate can reveal the position of buildings and hedgerows that

LAND LAID BARE
In Madagascar's central plateau, rapid deforestation has created a landscape of windswept grass and rapidly eroding soil.

have long since vanished. A piece of woodland may look as though it has remained unchanged for centuries, until a closer look shows that it contains a slightly different mixture of tree species from others around it, or ones that have grown into distinctive shapes. The reason may be that it was managed for a particular use in the past.

Some changes are bigger in scale, and are harder to explain. The strange behaviour of Lake Valencia, a large body of water in the coastal mountains of Venezuela, is one example. In 1800 the German naturalist and geographer Alexander von Humboldt reached the shore of the lake and discovered a curious fact: the water level had dropped at least 13 ft (4 m) in the previous 70 years, and was continuing to fall.

Twenty-five years later a leading French agricultural chemist, Jean Baptiste Boussingault, arrived at the same lake, and found that since Humboldt's visit it had mysteriously risen again. But in the succeeding years the trend went back into reverse, and by the 1970s the lake had plunged 55 ft (17 m) from the level Humboldt recorded – so far, in fact, that it now no longer empties into the river that once drained its waters into the sea. The lake has risen and fallen like a pointer in some vast measuring instrument. But what is it measuring? And is the change natural, or is it man-made?

Both Humboldt and Boussingault correctly deduced the answer. Lake Valencia lies in one of the most fertile valleys in the South American continent, and even in

1800 much of the land surrounding it had already been cleared for crops. By removing the forest around the lake, local farmers altered the conditions and climate in the valley. Over the decades the air became drier, less rain fell, and more water was diverted to irrigate fields. As a result the lake's level fell, first gradually and then with increasing speed. Without its normal flow of fresh water it also became more salty and more polluted, affecting water wildlife and the people who lived on its shores.

But what of Boussingault's observations, which showed the lake's level briefly rising once more? Boussingault himself provided the explanation. During the early 19th century the people of Venezuela – then a part of the Spanish Empire – were engaged in a war for independence and, in the ensuing disorder, agriculture fell into decline. As fields were abandoned, parts of the lakeside forest crept back and the lake regained some of the water it had lost. Thus the lake recorded the fortunes not only of the local farmers but also of the country in which they lived.

CAUSES AND EFFECTS

The story of this South American lake is by no means unique, although it is particularly well documented. As people have spread across the Earth, and as the human population has multiplied, there have been many changes like this. But until the birth of industrialisation, about three centuries ago, human

influence worked mainly on a local level. Now it is felt worldwide.

In the early 19th century most man-made changes could be read from the landscape and taken in with the sweep of an eye. By the 1980s the scale of change had altered, and something quite different was needed to perceive it. In 1984, tracing an orbit that took it high over the South Pole, the Nimbus-7 weather satellite gauged the depth of the ozone layer in the world's atmosphere, and then transmitted the data back to Earth. As the data for September, October and November came in, a startling fact became apparent. Over the entire continent of Antarctica, and in parts of the surrounding Southern Ocean, the ozone layer had become so thin that it had practically ceased to exist. Normally the ozone layer blocks potentially damaging ultraviolet light, but in the 1980s the shield failed, allowing potentially damaging radiation to reach the ground.

It is one thing to alter the lie of the land, or to change the

level of a lake. But changes in something as vast and deep as the Earth's atmosphere are developments of a far greater magnitude, and of far greater significance. The atmosphere protects and sustains life on Earth; if it changes in any way, the delicate balance between living things and their environment is almost certain to change too.

When meteorologists investigated the development of the Antarctic ozone hole, they probed a part of the environment which apparently lies far beyond human reach. Yet despite its apparent remoteness from the world we inhabit, the ozone hole is a man-made phenomenon, triggered by

few voices were raised about the long-term consequences. Today we know that the natural world is not an unlimited resource that can be tapped without restraint.

This realisation has come too late to halt the changes that have already taken place in the natural world, but it is altering the way that we treat our planet today. Across the globe, special reserves have been set up to protect natural landscapes, and in the far south an entire continent has been placed beyond the reach of development. At one time Antarctica's whales, seals and even penguins seemed destined to follow the fate of many other animals that have been hunted to the edge of extinction, and despite its severe climate its coal, oil and minerals attracted increasing interest. For the present, at least, that threat has been lifted. It is a fitting end to an extraordinary story. After emerging from obscurity to dominate and shape a planet, we have reached the point at which change itself has, in some cases, been brought to a halt. We have changed the Earth, but in the process we ourselves have changed as well.

RAPID RECOVERY *Two fur seals laze near the shore in South Georgia. Following a ban on hunting, this species recovered from just a few hundred animals in the 1930s to over a million today.*

volatile chemicals that were once thought to be harmless. It is proof that, in the modern industrial age, our reach extends across the whole of the Earth, and even high above it.

THE LAST WILDERNESS

Until well into the 20th century the natural world was seen as something to be tamed and exploited. To those with vivid imaginations, wildernesses were the haunt of predators such as bears and wolves, of poisonous insects and dangerous plants, and of deadly diseases. Compared with this, an orderly landscape of fields, farms and towns offered security and prosperity. Since those times cities have expanded, roads and railways have fanned out across continents, rivers have been dammed, and deserts turned into productive fields. In some places entire hillsides have been shaved away for the minerals they contain, and in others completely new coastlines have been created. Tropical rain forests – of the kind that Humboldt

encountered on his tour of South America – are rapidly shrinking, and many of the world's largest animals, such as the tiger and black rhinoceros, face an uncertain future.

However, while we now have the ability to create change on an unprecedented scale, we have also started to draw on some of the lessons of the past. When the world's first farmers set about transforming areas such as the arid plains of the Middle East, nobody understood the hidden dangers of irrigation, or the problem of eroding soil. Today these dangers are well known, even if they are not always easy to solve. When the world's first factories opened for business, there were no environmental assessments or regulations to prevent pollution. Now pollution control is an important part of any industrial process. And when, in the rush for the resources for industry, nature was over-exploited,

TEST-TUBE PLANTS *These seedlings have been grown from single cells taken from a parent plant. Techniques like tissue culture will have profound effects in years to come.*

A HUMAN REVOLUTION

1

RICH RETURNS *An Egyptian tomb painting shows grapes being gathered.*

OVER 3 MILLION YEARS AGO, MODERN MAN HAD YET TO EVOLVE, AND OUR DISTANT ANCESTORS LIVED IN THE SAVANNAHS OF AFRICA. COMPARED WITH OTHER ANIMALS OF THEIR TIME, OUR FOREBEARS WERE NOT OUTSTANDINGLY RESOURCEFUL; BUT DURING THE PROCESS OF EVOLUTION, SOMETHING HAPPENED THAT WOULD CHANGE THE EARTH: THESE APE-LIKE PRIMATES DEVELOPED INTO HUMANS, AND A SPECIES WITH INSIGHT WAS BORN. SCIENTISTS HAVE YET TO AGREE WHY THIS HAPPENED, BUT THE CONSEQUENCES WERE DRAMATIC. INSTEAD OF REMAINING IN AFRICA, HUMANS FANNED OUT INTO EVERY CONTINENT EXCEPT ANTARCTICA, DEVELOPING THE TECHNOLOGIES AND SKILLS THAT WOULD EVENTUALLY MAKE THE ENTIRE PLANET THEIR HOME.

CROWDED *Humans put growing pressure on nature's resources.*

THE WORLD WITHOUT MAN

In nature, everything must struggle to survive. Our early forebears succeeded in this struggle, but they might well have failed. In a world without the presence of humans, the Earth would have evolved in a very different way.

What would the Earth be like if humans did not exist? The answer is: a planet very different from our own. It would still carry a teeming cargo of living things in a yearly orbit around the Sun, but that cargo would lack one element of overwhelming significance: a form of life able to dominate all others, capable both of transforming its own planet and of reaching out far beyond. If, by some strange miracle, visitors from our planet reached this imaginary unpeopled one, many of the differences would be immediately apparent. Many places in today's Earth have open landscapes only because we have made them that way. In an Earth without human life, these would still be in their original forms. There would be no fields, no roads and no cities, and most of the ground that these now occupy would be covered by dense forest, natural grasslands or the low-growing plants that flourish on level shores. The only sounds would be ones heard on Earth long before

human life first appeared: the whisper of wind through grass or trees, the rumble of thunder from distant storms, the rhythmic breaking of endless waves, and the calls of wild animals on the empty air.

For although devoid of people, this world would be far from deserted. Even a brief exploration would reveal a breadth of wildlife that we would find fascinating, but also distinctly daunting. In North America, for example, the open grasslands would be cropped by vast herds of buffalo, led by males weighing up to a ton. Wolves would be an everyday sight throughout the Northern Hemisphere. Tigers could be encountered anywhere from the shores of the Black Sea to South-east Asia, and lions would be common throughout Africa, southern Asia, the Middle East and parts of southern Europe. Across all the continents, a vast army of lesser predators – many of them rare in our own world – would take their toll of Earth's abundant life.

On remote islands the sea would act as a protective barrier, as it did on the real Earth before the days of boats and ships, shielding unusual life-forms from those that had evolved elsewhere. On some islands, flightless birds taller and heavier than a man would stare blankly at any intruders that somehow reached their shores, while on others lumbering tortoises – more than 5 ft (1.5 m)

long – would lift their scaly heads at the sound of approaching feet, before crawling away in their continuing search for food. In these island fastnesses the arrival of visitors would be greeted not with fear or even surprise, but with complete incomprehension.

THE UNTAPPED EARTH

In every kind of habitat, from forests to oceanic islands, life would be different without human influence. But in its physical make-up as well, some features of this imaginary Earth would be quite unlike our own.

The most obvious difference would be the lack of structures built by human hands:

no power lines or dams, no quarries or mines, no pipelines or radio masts – in fact, nothing artificial at all. Their absence would be immediately apparent, but after some days in this kind of world a visitor from our own planet might notice something else. Without humans, mathematically precise objects of any kind – the hallmark of any technological civilisation – would be almost completely non-existent. Nature often crafts straight lines and flat surfaces on the smallest of scales, for example in crystals, but on a more everyday scale linear shapes are rare.

On this uninhabited Earth, a visiting geologist or surveyor would find an abundance

of untapped wealth. Some of the materials that we take for granted, such as wood, would be present in almost overwhelming profusion, but on a planet that had not been scoured by generations of prospectors, other treasures, too, would be in store. In some places outcrops of coal would crumble into scattered blocks, untouched as a source of fuel, and in others pools of sticky oil, oozing up from natural reservoirs underground, would form traps for unwary animals. A painstaking search of the surface might eventually reveal small deposits of precious metals, including copper, silver and gold, though many of the metals that we use in everyday life – such as aluminium and zinc – would prove impossible to find. These elements are extremely abundant in the planet's crust, but they are always locked up in combination with other substances. Only human ingenuity has parted their chemical embrace.

This parallel Earth would lack many of the complex compounds that we use in daily life, including all our plastics and most of our solvents. It would also be without some forms of matter that even in our populated world are outstandingly rare. In its natural state, the Earth has just over 90 chemical elements, of which about 25 are used by living things. Fleetingly, and in tiny quantities, nuclear physicists have managed to bring into being over a dozen more. Most exist for just a few millionths of a second before radioactive decay whittles them away and transforms them into something else.

CHANGES FROM OUTSIDE

The journey back from this imaginary world would be one of striking contrasts. Some parts of our planet – mainly those that are very dry or cold – are still thinly populated, and look much as they would have done before the human race appeared. Others are home to billions of people, and have been

POINT OF IMPACT
The origin of Arizona's meteor crater (above) was not fully explained until the early 20th century. This fragment of an iron-bearing meteorite (left) gets its blue colour from a mineral called olivine.

altered in far-reaching ways. But even if we did not exist to bring about these alterations, the Earth would not stay unchanged for long. Instead, natural processes would continue to reshape its surface, creating new opportunities for some living things and drastically limiting the prospects for others.

Most of these processes work slowly, but some take place almost in the blinking of an eye, occasionally with devastating results. One of the most impressive pieces of evidence for this kind of event can be seen in

the desert of Arizona, halfway between the towns of Flagstaff and Winslow. Here, about 22 000 years ago, a meteorite survived its passage through the atmosphere and struck the ground with the force of a 4 megaton bomb. The result was a huge impact crater, which even today – after thousands of years of erosion – is nearly ³/₄ mile (1.2 km) across and over 650 ft (200 m) deep. Given a fleet of bulldozers and mechanical diggers, it would be possible to excavate a hole of this size, but the work would take several years.

Geologists calculate that the crater formed in less than a minute.

In other parts of the world far bigger craters, also caused by the impact of chunks of rock smashing in from space, punctuate the Earth's surface. In eastern Quebec, for example, an unusual ring-shaped lake marks the perimeter of Manicouagan crater which, at 47 miles (75 km) wide, can be clearly seen only from space. The energy that formed this structure was probably as great as the energy in all the earthquakes that take place in a thousand years, and the shock waves it produced would have thrown a dust-cloud around the whole Earth.

Fortunately for us, the impact that formed the Manicouagan crater took place over 200 million years ago, long before humans had evolved. Although the Arizona impact occurred much more recently, it is unlikely – although just possible – that humans were present to witness the moment of catastrophe. However, the destructive power of nature has been demonstrated nearer to our own times. Mount Tambora, a volcano on an island in modern-day Indonesia, exploded so violently during an eruption in 1815 that more than 40 cu miles (170 km³) of ash were thrown into the air. In 1908 an unidentified explosion blasted the remote Tunguska valley in central Siberia, destroying a huge area of coniferous forest. The Tambora eruption caused many deaths, and if the Tunguska explosion had occurred in a more populous part of the world the consequences would have been impossible to imagine.

For nearly two decades, wars and revolutions prevented the Tunguska impact from being fully investigated. When a team of Soviet scientists eventually managed to reach the region in 1927, they found hundreds of square miles of flattened forest, but no signs of a crater. The trees radiated outwards from a central point, but, remarkably, those at the centre itself were still standing, although completely stripped of their branches. The most likely explanation for this pattern of devastation is that a rocky asteroid exploded before it struck the ground, leaving only microscopic debris. This, at least, is the current conjecture: the real truth may never be known.

A SHIFTING SURFACE

In the last 1000 years more than 500 major volcanic eruptions have been recorded, and about 100 impact craters – some even larger than the one at Manicouagan – have been identified by geologists as coming from the same period. But, despite the immense forces behind these eruptions and cosmic collisions, they play only a minor part in reshaping the surface of the Earth. Far more important are processes that

work slowly, methodically and without any great drama. Together they add up to a more gigantic whole.

To visualise how these changes can build up, imagine sitting on a river bank somewhere far from the nearest human habitation. The tranquil sound of flowing water fills the air, but apart from this there is silence. Suddenly, from the opposite bank, the stillness is broken by a dull clattering of rocks and soil, followed by a succession of splashes as the debris tumbles into the water below. Undercut by the current, a few inches of the bank have been eaten away.

The impact of this miniature landslide is minor, and, at any one point on the

river bank, weeks or months could go by before it is repeated. But if time speeds up so that a whole year passes in the course of each minute, and if our viewpoint shifts from the river's edge to a point high above it, things look very different.

Far from being a fixed feature of the landscape, the river now seems to behave like a living thing. Restlessly responding to the pull of gravity, it digs away at its banks, and snakes from side to side. During this process it constantly exaggerates its own curves, and then cuts through them once more as if deciding that its original course was perhaps the best. This creates a network of abandoned channels that record

A SLICE THROUGH TIME

Moving water is one of the most important agents of change on the surface of the Earth. It erodes rock with slow but unremitting thoroughness, and then sweeps away the debris to build it up somewhere else. For the most part the results of this ceaseless reworking of the crust are hidden beneath our feet, but there are some places where they are revealed in a truly breathtaking way. Foremost among them is the Grand Canyon, where nearly 2 billion years of geological history are visible at a glance.

Despite being up to a mile (1.6 km) deep and over 250 miles (400 km) long, the canyon itself is probably no more than 30 million years old, which makes it a relatively youthful feature. However, the rocks that form its cliffs and bluffs are far more ancient, and date back to times long before the present North American continent even existed.

WRITTEN IN ROCK *The setting sun throws into sharp relief the contrasting strata of sedimentary rock on the Grand Canyon's North Rim.*

As the land has been lifted up the Colorado River has cut into it like a liquid saw. It has worked its way through layer upon layer of these rocks, and has exposed parallel strata of sandstone, shale and limestone, whose colours and textures contrast like layers in an immense and crumbling sandwich. These strata began life as sediments in long-vanished seas, but the canyon's deepest rock, which cradles the current riverbed, has a different

history. Known as schist, it was created when heat and pressure transformed existing layers of lava and sediment, giving them a crystalline structure that is harder for water to erode.

Until the 1930s the Colorado was one of the world's siltiest rivers, gouging out about 400 000 tons of sediment every day. Dams have since stemmed its flow, putting a brake on the process that has carved out the most spectacular chasm on Earth.

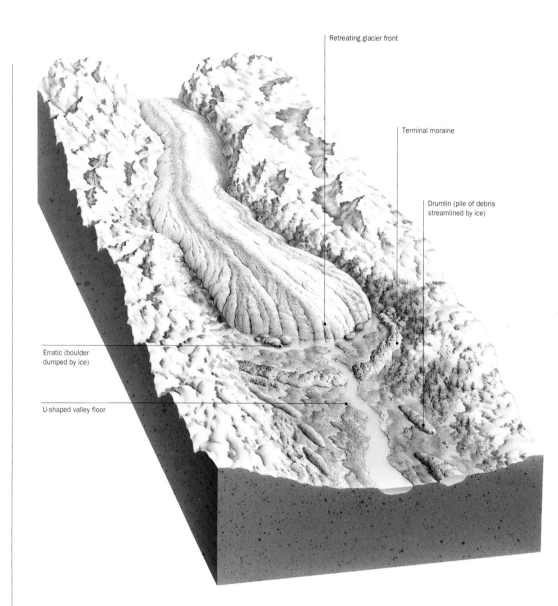

Retreating glacier front

Terminal moraine

Drumlin (pile of debris streamlined by ice)

Erratic (boulder dumped by ice)

U-shaped valley floor

EROSION BY ICE As a glacier flows downhill, it piles up banks of debris in moraines. Like tidemarks on a beach, moraines show how far the ice once extended.

ice front ends. But on one of these alpine excursions a glaciologist pointed out to him that moraines not only exist where the ice comes to a halt, some of them can also be seen in valleys far below the ice front, disguised by grass and trees but clearly visible to the expert eye.

How do these rocks manage to travel so far downhill? And, still more puzzling, how do barn-sized boulders manage to reach places even farther afield, where they sit still more incongruously in the landscape? The answer, Agassiz was forced to conclude, is that glaciers are no more fixed in place than rivers or coastlines. Instead, they advance and retreat, shovelling rocks before them, and then abandoning them when they pull back.

From his observations in the Alps, Agassiz became the most prominent supporter of a radical new theory, which proposed that large parts of the Northern Hemisphere had once been covered by sheets of glacial ice. In the 1830s many scientists found this idea unbelievable and even outrageous, but as the decades went by an overwhelming body of evidence began to back it up.

Armed with this theory, geologists have been able to make sense of many surface features that were formerly mysterious. Streamlined hills called 'drumlins', all pointing in the same direction, bear the rounded contours imposed by the ice's crushing weight. Snaking banks of rock and gravel called 'eskers' reveal the path of streams that once flowed beneath the ice, while 'striations' and jagged 'chatter' marks

the river's wanderings, like marks on a blackboard which have been imperfectly rubbed out.

Similarly accelerated, many other physical features are not as permanent or as static as they seem. Lakes fill with sediment and shrink like puddles drying in the sunshine. Shores advance and recede as the sea builds up some areas and erodes others. On mountains, exposed rocks are shattered by frost and undermined by rain, and in deserts, sand dunes scurry across the ground like flocks of shapeless animals, spurred onwards by the prevailing wind. Without any help from us, the fabric of the planet is on the move.

Human lifetimes are so brief that, even with the benefit of hindsight, these cumulative changes are difficult to perceive. We

notice when houses topple over collapsing cliffs, when rivers burst their banks, or when roads are smothered by sand, but long-term changes are much less obvious. However, each of them leaves its own telltale signature, and, for those who know what to look for, the Earth's changing past is written into its surface.

READING THE EARTH

In the summer of 1836 this fact struck the Swiss-American naturalist Louis Agassiz as he was exploring the glaciers of the Alps. Agassiz was aware that glaciers flow downhill – a fact that had only recently been established – and he knew that they gouge out rocks as they move. He had also studied the piles of boulders, called terminal moraines, that are dropped by glaciers where their show where the ice picked up sharp stones and gouged them against the bedrock below. The last Ice Age came to an end roughly 10 000 years ago, but its imprint remains.

AGEING THE EARTH

Ice Age theory explains many of the features seen in mountains and in high latitudes, but it does not explain how their rocks came to be laid down in the first place. For early geologists grappling with this perplexing question, a crucial difficulty was time. Until the 18th century the Earth was thought to be just a few thousand years old, which meant that gradual processes, such as erosion, could have had only a minor impact on its surface. The belief was widely held that the Earth had been shaped not by piecemeal events but by a series of catastrophes such as floods and earthquakes.

In the mid 18th century the French naturalist Count Georges-Louis Buffon, a colourful figure whose varied interests ranged from botany to astronomy, carried out an experiment designed to put the Earth's age to the test. It was generally agreed that the Earth had formed from a molten state, so Buffon tried to gauge how long it might have taken to cool. He heated

MOULDED HILL *Contour ploughing (right) highlights a drumlin, or hummock made of glacial debris. Some drumlins occur in large clusters called swarms. Dumped by retreating ice (below), a giant boulder or erratic sits incongruously on the shores of an Alaskan lake.*

some metal spheres in a furnace until they had almost reached melting point, recorded how quickly their temperatures fell, and then extrapolated the results to the planet as a whole. Although Buffon's physics were a little shaky, his results – an age of at least 75 000 years – caused a great amount of interest, and made some geologists question their earlier assumptions. If Buffon was on

ROCK CYCLE *This sandstone cliff (left) is made of vast numbers of tiny sand grains, eroded from rocks that existed long ago. In turn it will be eroded, helping to form the rocks of the future. The heat of the Earth's core generates upwelling lava (opposite). When it solidifies, lava creates igneous rocks such as basalt and granite.*

the right track, the Earth might be old enough to allow apparently minor processes to add up, producing major effects.

Strangely enough, this idea was not a new one. In his *Book of Minerals* the 11th-century Arab scholar Ibn Sina, known in the West as Avicenna, described how rivers cause erosion and sweep sediment into the seas, and suggested that this sediment might eventually form new rocks. Hundreds of years later, when Buffon carried out his work, these far-sighted ideas began to gain a new audience.

ENERGY FOR CHANGE

By human standards 75 000 years is a vast sweep of time, but in a world of gradual changes it is still extremely brief. In a period of this length, enough sediment could be washed down rivers and into the seas to form a layer of rock at most several inches thick, but in some parts of the world sedimentary layers reach down for more than half a mile (0.8 km). Even in Buffon's time it was apparent that the figures did not add up.

In the 19th century the British physicist William Thomson, who was an expert on the movement of heat, followed in Buffon's

---

THE FURNACE WITHIN

The temperature of the Earth's molten core is estimated to be at least 5000°C (9000°F), but fortunately for living things this heat is held in by the solid crust and leaks out extremely slowly. The rate of heat loss varies hugely from place to place. It is highest along the margins of tectonic plates, where new crust is being created or destroyed, and lowest in the centre of continents, where the crust is oldest. On average it would take all the energy lost through 11 000 sq ft (1000 m²) of ground to power a standard light bulb.

---

footsteps and made his own estimate of the Earth's age. Unlike Buffon, he started by calculating the current temperature of its core and then worked out how long it would have taken its former heat to leak away. This produced a figure of about 100 million years – vastly greater than Buffon's 75 000 years, but still a fairly meagre amount of time for a vast amount of change.

As it turned out, Thomson's calculations were almost flawless, but he lacked one vital

piece of information that would have produced a very different result. As well as losing heat, the Earth also generates it by the radioactive decay of elements such as uranium – a fact that was completely unknown when Thomson did his work. As a result, the planet is cooling much more slowly than would otherwise be the case. Uranium atoms decay to produce atoms of lead, and they do this at a slow, exact and unvarying rate. By measuring the ratio of uranium to lead, it is possible to date rocks – and by extension the Earth itself – with remarkable precision.

Through radiometric dating, the oldest rocks on the Earth so far tested are known to have existed for approximately 3.8 billion years. The planet itself is thought to have formed about 800 million years before that, at roughly the same time as the Moon; but while the Moon is cold and inert, our world is still awash with energy. From the moment that the first solid surface appeared, the Earth's inner heat has kept the continents in motion, while the Sun's radiant energy has powered the wind and rain that have worn them away.

In the very distant future the Earth's inner fire will fade, and this process will eventually come to a halt. Until then, everything on the planet's surface – including the highest mountains, the deepest ocean trenches and the continents themselves – will remain transient features, awaiting the time when they are changed beyond recognition, or swept away.

A SPECIES APART

Why are humans so successful? What features separate us from all other forms of life, and how did they first appear? The full answers to these questions may never be known, but they involve two key developments: technology and language.

All living things alter the world around them. A sea urchin grinding its spines against a rock creates a hollow cup that protects it from the powerful currents in its seashore home. A tree, probing downwards with its roots, draws up minerals from layers deep below, while a beaver chews through saplings and then floats them into position behind a dam, helping to hold back a lake that will protect both itself and its young. Even after death, living things can still influence their surroundings. Dead leaves create soil that blankets the ground. The bones of dead whales or seals, washed up on a beach, provide both shelter and fertiliser for low-growing plants.

From observations like these, it seems only natural that our most distant ancestors must also have had some effect on their surroundings. Like other primates in Africa, they would have picked up objects and carried them from place to place. By scattering seeds they would have helped

EARLY TOOLMAKERS
Our distant ancestors made tools from stone (left). The tools had many uses including breaking bones (middle) and cutting up animals that they had killed (right)

HANDY TRICK *The woodpecker finch is an adept tool-user, but unlike a human it has no understanding of how tools work or can be improved.*

some plants to spread, and by catching small animals they would have influenced the numbers of some of the species around them. But until less than 3 million years ago their impact was no greater than that of any other animal; it simply fitted into the much greater pattern of living things interacting with their environment. And that is how it might have remained. But then, at some unknown point in the past, something happened: they learnt to make tools.

TOOLS AND TOOLMAKERS

Many decades ago, biologists credited the human family – which includes ourselves and our immediate ancestors – with being the only forms of life to use tools. It was a bold claim and, as it turned out, a false one. We are, in fact, far from unique in using tools, and the more closely biologists observe animal behaviour, the more examples of tool use crop up. Tool-users include not only non-human primates, such as monkeys and chimpanzees, but also other mammals, such as sea otters and mongooses, and many kinds of birds, including woodpecker finches, fly-catchers, parrots, herons, jays and crows. Their tools are usually no more than stones or pieces of stick, but they are still effective aids for the job they have to do, which usually involves getting at food.

Following this realisation, scientific opinion converged on the idea that tool-making, rather than tool use, sets us apart from other animals. But here again we are not alone. Some animals simply pick their tools off the ground, but others shape or process them in some way. For example, woodpecker finches create tools by snapping spines off cacti, while chimpanzees strip the leaves and bark off twigs. In one

remarkable instance of tool-making, a captive blue jay (*Cyanocitta cristata*) was even seen to tear up strips of news-paper, which it then used to pull food towards it. Clearly, toolmaking is not uniquely human either.

So what is it that sets us and our ancestors apart from other tool-users and toolmakers? The answer is that, to begin with, there was very little. Then, as human evolution progressed, the line that led to humans crossed a crucial threshold. Instead of turning out implements in a stereotyped way, like a bird snapping off a cactus spine, our ancestors gradually developed an understanding of what it was they wanted to create, and how it would be used. In short, they acquired insight.

Establishing when this happened and how is difficult. *Australopithecus afarensis*, a human ancestor about the size of a chimpanzee, lived some 4 million years ago. It walked upright, and would certainly have used its hands to collect food, and perhaps other useful objects such as sticks and stones.

FISHING FOR FOOD *This chimp is 'fishing' for termites in a broken tree trunk. It is using a stick that it has selected and shaped for this purpose.*

It may have been a tool-user and perhaps even a toolmaker, but so far no evidence for this has been found. The most ancient surviving tools made by our ancestors are shattered stone pebbles, found at Hadar in Ethiopia. They are probably about 2.5 million years old, and were made by *Homo habilis*, or 'Handy Man'. Although crude, they already reveal evidence of both foresight and insight: the best stone was available only in a limited number of locations, and *Homo habilis* clearly knew where to get it.

TOOLS TO CHANGE THE WORLD

For toolmaking animals, the business of 'manufacturing' a tool is usually quick and straightforward. A woodpecker finch, for example, simply has to break off a twig or cactus spine, and its tool is then ready for use. A chimpanzee sometimes takes more trouble in making its tools – for example, by carefully selecting a suitable stick and peeling off its bark – but even this is the work of just a few minutes. Once the tool has been used to obtain food, it is often abandoned.

At first, human toolmaking would have been equally rapid and rudimentary, and the results equally disposable. Oldowan tools – the type made by *Homo habilis*, and named after Olduvai Gorge in Tanzania – were usually core tools, which means that they consisted of the cores of rocks or pebbles that had been struck to give them an edge. The most primitive of these tools, which date back more than 2 million years, would have been crafted with just a handful of blows, and if lost or abandoned would have been relatively easy to replace.

By comparison, Acheulian tools – the type made by *Homo erectus*, and in use until about 100 000 years ago – needed more time, work and forethought to produce. Some of these tools consisted of cores, but many were based on flakes chipped off cores, which were then shaped to produce cutters, scrapers or awls for piercing animal skins. By about 30 000 years ago, when Cro-Magnon people – early *Homo sapiens* – lived throughout Europe, tools were often beautifully finished, and the work that went into them strongly

suggests that they were objects of considerable worth.

In economic terms, this was an important change. As tools began to develop a high inherent value, they joined a small group of other objects, such as animal skins and spear-shafts, that were worth carrying from place to place. This marked the start of the human penchant for possessions.

FINELY CRAFTED *A typical Acheulian hand-axe (right) was a precision-made instrument, cut to an exact size. Left: Some tools were shaped for specific purposes (left to right) a stone core, a side scraper and a point.*

By about 2 million years ago these proto-humans had advanced to the point where they had a range of distinct tools. We know only the ones that have survived – which are made of stone – but it is quite likely that they had other ones made of organic materials, such as wood, bark and horn, which rarely survive once they are buried. Using these implements, *Homo habilis* could cut through hides, slice up meat and break open bones. The age of technology had begun.

THE IMPACT OF EARLY TOOLS

Unlike *Australopithecus*, which was largely a plant-eater, *Homo habilis* clearly turned its toolmaking abilities against the animals around it. Piles of bones in ancient campsites show that it fed on animals as large as elephants and giraffes, although it is not clear whether *Homo habilis* killed all of its food itself – some of it may have been obtained by scavenging the kills of other predators. It is also possible that *Homo habilis* built rudimentary shelters, although the evidence for this is scant. However, with a total population of perhaps just 1 million individuals, spread out across a large part of Africa, these early humans can have had only a limited effect on their surroundings and on the other animals that shared their habitat.

Homo habilis died out about 1.5 million years ago, and by that time another tool-maker had emerged in Africa. Unlike *Homo habilis*, *Homo erectus* produced a more extensive range of stone tools, some of which were fashioned by dozens of carefully aimed blows. In places – for example, Olorgesailie in Kenya – the ground is littered with thousands of teardrop-shaped hand-axes, each weighing more than 1 lb (0.5 kg) and shaped to fit snugly into the palm of a hand. With tools like this, *Homo erectus* became a formidable predator, and was a threat to the largest mammals on the African plains.

Looking back through the telescope of time, it is easy to form the impression that the art of toolmaking took off as *Homo habilis* gave way to its bigger and more intelligent successor. But the most remarkable thing about progress in toolmaking is that –

for most of human history – it seems to have been staggeringly slow. Today we are used to constant innovation, but in the days of our distant ancestors innovation was extraordinarily rare. At Olduvai Gorge in Tanzania, excavations have revealed that there were about six basic types of stone tool 2 million years ago, when *Homo habilis* still thrived, and a million years later, when *Homo erectus* existed in the area, the total tool count had risen to just ten. Without allowing for tools that may have escaped discovery, this suggests an innovation rate in this area of one new tool every 250 000 years. Our distant ancestors may have had some insight into the process of making tools, but coming up with new designs was not foremost among their talents.

Despite this plodding rate of change, these early humans had reached the point where they could exert a significant impact on their surroundings. *Homo erectus* certainly built shelters, and may have played the role of a key predator in some places, limiting the numbers of some of its prey. It has also been suggested that it was responsible for the extinction of many of Africa's large mammals, such as the giant gelada baboon (*Theropithecus oswaldi*), which weighed almost twice as much as a modern gorilla. But as nomadic bands of *Homo erectus* spread throughout Africa, and eventually into Europe and Asia, tools were not their only means of altering the world around them: they had also learnt to use fire.

FIRE-USERS AND FIRE-MAKERS

There can be few better places for learning about fire than the grasslands of East Africa, where *Homo erectus* once ranged. Each year brings a long dry season, which comes to an end when

FOOD FROM THE FLAMES *In the grasslands of tropical Africa fires quickly attract birds that feed on animal casualties.*

thunderstorms move in from the Indian Ocean. As the clouds roll across the open plains and escarpments, the air is charged with electricity, and lightning flickers over the parched landscape, heralding the rain that is about to begin. If a lightning bolt hits a dead tree or the dry grass in open ground, fire is the almost certain outcome.

These fires are frequent events in seasonally dry habitats all over the world, and the plants that grow in such places have little difficulty coping with them. The flames quickly burn off dead grass, but the temperature of the ground rarely rises above 100°C (212°F), so the grass roots are left undamaged. Having exhausted their meagre supply of fuel, the flames soon move on. They leave behind a blackened landscape, cloaked in mineral-rich ash that soaks into the soil with the rain, acting as a fertiliser.

In a light wind, the flame front moves at the speed of a leisurely walk, but even so it is too fast for some small animals in its path. Grasshoppers and beetles often fail to get airborne in time to escape, and their wings crackle into flame as the fire overtakes them.

Small lizards are sometimes engulfed as they run towards the safety of rocky crevices, and snakes may find themselves hemmed in by the wall of heat. In Africa, animals like the kori bustard (*Ardeotis kori*) and marabou stork (*Leptoptilos crumeniferus*) are quickly attracted by the prospect of a fireside meal.

It is almost certain that early humans were attracted to the flames for much the same reason, but instead of just standing by, like bustards and storks, they learned to interfere with fire, turning it to their advantage. For example, they would have noticed that the cooked flesh of burnt animals was easier to eat than raw meat, so it is quite likely that they would have dropped dead animals into the flames, to cook them as the fire passed. They would have seen fires fail for lack of fuel, and at some point would have learned to pile wood and leaves onto them in order to feed the flames. During this process they may well have found themselves clutching one end of a burning stick. Once their natural fear had subsided, they would have discovered that this was a safe way to carry fire from place

to place. All of these events would have happened on many occasions in many different locations, as *Homo erectus* found that fire could be exploited and controlled.

The image of these early people, gathered around a flickering campfire at the end of the day, has a deep resonance for us, and marks them out as undeniably human. However, opinions differ widely as to when scenes like this might have first occurred. Estimates vary from about 1.5 million to about 400 000 years ago, with the earliest firesides being scattered across the whole of the Old World, from southern Africa to China. In fact, the earliest evidence of people making fire – as opposed to carrying and maintaining it – is remarkably recent. It dates back only 15 000 years, to a time when *Homo erectus* was long extinct and had been supplanted by our own species, *Homo sapiens*.

CULTURAL EXCHANGE

With their knowledge of tools and fire, our immediate predecessors in the human family – *Homo erectus* – had embarked on a path that increasingly diverged from that of other primates. They managed to spread to

a remarkably wide range of habitats, from tropical grassland to icy tundra, and adapted to a variety of different diets, based on what they could find in these very different surroundings. They were the ultimate opportunists, quick to exploit anything that might provide a meal.

As these early humans evolved, they changed shape. Their jaws became shorter from back to front, and their brains increased in size. Changes like these occurred through natural selection, which drives evolution throughout the living world. Natural selection works when living things reproduce, by favouring useful genes at the expense of others. But as well as passing on genes, these early humans also began to hand on something else – knowledge gained from their own experience.

In the living world all animals respond to their environment, and all but the simplest learn new forms of behaviour as time goes by. For example, when a bird first leaves the nest it does not know to avoid butterflies that taste unpleasant. However, after trying to eat one, it will quickly learn not to attack that species again. This kind

of behaviour is an acquired characteristic. Acquired characteristics do not alter an individual's genes, so the behaviour will not be passed on when the bird eventually breeds. As a result, unless the birds learn from their parents, each generation has to learn by making the same mistake.

THE FIRST FIRES

Several places have been claimed as the site of the first man-made campfires. One of the oldest is at Swartkrans cave in southern Africa, where fossil bones of antelopes and other animals, dating back over 1 million years, seem to have been subjected to much higher temperatures than are found in natural fires. However, no hominid fossils have been definitely linked with these remains, so the origin of the fire remains in doubt.

However, behaviour can be passed on through forms of learning that sidestep genes. If a chimpanzee discovers a new way of turning a stick into a tool, others may then copy it. If the copying process continues as different generations succeed each other, the new trick will be reproduced, and it will quickly spread. The chimps do not need genes to tell them how to make tools, just as we do not need genes to tell us how to use computers. They simply need genes that make them good at watching and at learning from what they see.

This way of passing on behaviour is called cultural exchange. Compared with many other animals, primates spend a good deal of time copying each other – as the expression to 'ape' something indicates. However, for all their copying, this cultural exchange is still quite limited. Yet in the line leading to modern humans, who first appeared about 200 000 years ago, cultural

LEARNING CURVE *A long childhood allows apes and humans to learn behaviour from their parents. In simpler animals, behaviour is controlled largely by instinct.*

exchange developed into a phenomenally powerful force – all the more so because our ancestors learnt to communicate through speech. They were able to pass on skills not only by example but also by language.

The effect of this cultural exchange is difficult to exaggerate. A cascade of knowledge started to flow from one generation to the next, helping humans to make the best use of their resources, and to make sense of the world around them. That cascade – in a hugely magnified form – continues today.

THE IMPACT OF HUNTER-GATHERERS

Palaeontologists are far from agreed about how our own species came into being. We may have arisen more or less simultaneously in Africa, Asia and Europe from earlier human species – an idea known as the multiregional hypothesis – or we may be descended from a fairly small group of *Homo sapiens* that evolved in Africa from an earlier human species and spread out from there to Europe and Asia, supplanting populations of *Homo erectus* and the like. Whichever of these two theories proves to be correct, one thing is not in doubt: at least 100 000 years ago humans already had an unprecedented ability to interfere with the natural world.

At this stage in human evolution, people existed entirely by hunting and by gathering whatever else they could find. It is difficult to assess how these hunter-gatherers affected their surroundings, because most of the evidence has faded long ago, but one way to find out is to look at hunter-gatherers that exist today, and ones that lived in the recent past. One thing is clear: even this primordial lifestyle can alter the natural balances between other living things. A study of the Aleutian Islands off Alaska has shown that native Aleuts, living between 2500 and 500 years ago, often had a severe effect on the region's sea otters, which they hunted for food and fur. In some places the sea otter was brought to the brink of extinction, and this removal of a single species triggered radical changes in the environment as a whole.

In the Aleutian Islands, sea otters often eat sea urchins. These live close to the shore, where they creep over submerged rocks, scraping away seaweeds and other algae with their pincer-like jaws. When the sea otters were removed, the sea urchins lost their worst enemies, and their numbers rocketed. They grazed the rocks so thoroughly that large seaweeds were unable to

FEELING THE PINCH *Hunter-gatherers are often thought to have little effect on wildlife. The history of the sea otter suggests that this is not always the case.*

become established, and without seaweeds there were fewer inshore fish. Without the fish, seals had little to eat, so their numbers slumped as well.

This kind of effect has been used by some scientists to explain the abrupt disappearance of many large land animals at several points during the history of human evolution. One involved the extinction of large African mammals during the time of *Homo erectus*, but a much more recent example

OUT OF THE WILD

The history of the horse illustrates the way that humans can shape the fate of other forms of life. Originally horses were hunted for food. Piles of bones – such as the one discovered in the village of Solutré in central France – show that our ancestors were adept at cornering and killing these animals, despite their speed and wariness. However, as time went by, the relationship between horses and humans changed. People began to follow horse herds more closely, and at some point – probably in Central Asia about 6000 years ago, but perhaps earlier in Ice Age Europe – domesticated horses first appeared.

Domestication marked the beginning of a great change for the species, as the remaining wild horses came under greater and greater pressure from hunting, habitat change, and competition for food from their domesticated counterparts. Within historical times there were three wild subspecies – the forest horse of Central Europe (*Equus caballus sylvaticus*), the tarpan of southern Russia (*Equus caballus gmelini*), and Przewalski's horse (*Equus caballus przewalskii*) of Central Asia. The first two died out in the 18th and 19th centuries, and the third was last seen in the wild in Mongolia in 1968, although several hundred survive in captive herds.

Meanwhile the domesticated horse has gone from strength to strength. There are now at least 180 different breeds, and the world's domesticated horse population runs to many millions, despite falling back in the 20th century following the development of the petrol engine. In a final twist to the horse's story, some of these domesticated animals have spread to new parts of the world, where they have escaped into the wild. These feral horses include the mustangs of North America, which have existed since Europeans brought horses to the continent, and the wild horses of Australia, which are also the descendants of animals that arrived with European settlers.

GREAT ESCAPE *North America's 'wild' horses have only existed since the 16th century, when domesticated horses escaped or were set free.*

occurred about 12 000 years ago in North America. Two-thirds of North America's large mammals disappeared in less than 1000 years, a catastrophic collapse that has puzzled palaeontologists for decades.

This wave of extinctions may have been the result of sudden change in climate, but similar changes in the past had nothing like such serious effects. A more likely explanation is that it had something to do with hunter-gatherers, moving south after reaching North America from Asia. These new arrivals would certainly have had the technology to kill a wide range of species, but even if they concentrated on hunting a few key animals, which were already under pressure from climatic change, they may well have reduced the food supply on which other animals depended. This could have created a 'knock-on' effect, killing off animals such as mastodons and ground sloths.

THE UPS AND DOWNS OF FARMING

For more than nine-tenths of our history as a species, humans have lived solely on what nature provides. Hunter-gatherers fanned out across all the continents except Antarctica, and successfully exploited an extraordinary assortment of foods, ranging from football-sized fungi that grow on trees in Tierra del Fuego to insect grubs in the rain forests of Central Africa. In good conditions, hunter-gatherers could collect all they needed during just 20 hours of foraging a week. Some were nomadic, but others – particularly those who lived on coasts – formed settled communities that rank among the world's first urban areas.

Given the successful record of this way of life, what happened about 10 000 years ago seems all the more remarkable, and all the more difficult to explain. In several regions across the world – most notably the Middle East – people gave up an existence based on collecting food, and took to one that involved growing it instead.

The origins of farming, like the origins of modern humans, are a source of long-running debate. It is easy enough to guess how people might have created domesticated plants – by collecting and planting species that were useful – but why they took up farming in the first place is a more perplexing question. With the help

of machinery or domesticated animals, to-day's farms can be hugely productive, but the first farmers had none of these aids. They had to raise their crops entirely through their own labours; they also had to guard their fields, build safe shelters for their harvest, and devote time to making their grain ready to eat. As a way of life it had many disadvantages.

Some experts believe that farming was spurred on by population growth. Because farming can generate more food than each farmer needs, it would have allowed more people to survive on a given area of land. But others see it more as a last resort, forced on people by a changing climate. According to this theory, which is based on research carried out in the Middle East, cereals only came into prominence after a cold, dry spell that lasted from about 11 500 years ago to 10 500 years ago. Once people had been forced to cultivate these plants – this theory suggests – they would have come to appreciate the advantages of growing food that could be stored against hard times ahead. For hunter-gatherers, building up this kind of reserve was not an option.

THE SETTLED LIFE

We may never know precisely why farming started, but in less than three-millionths of the total span of life on Earth it has created the greatest biological changes the

LIVING LARDER *For hunter-gatherers, small animals like honeypot ants provide a useful source of food.*

world has ever seen. Unlike hunting and gathering, this new lifestyle does not simply dabble with the natural world; it often changes it beyond recognition.

These changes work on many levels, from the landscape itself to the things that are raised within it. In all ecosystems, nature weeds out species that are less effective at survival, creating opportunities for ones that are better at exploiting local conditions. As a result there is constant change. However, this kind of change takes place extremely slowly, because after millennia of struggling for supremacy the competing species are already extremely well matched.

When humans enter the picture, things are very different. Where natural selection is finely balanced, selection by humans is brisk, simple and ruthless. Farmers try to remove all competing forms of life – weeds and pests – and constantly select the best examples of each crop or animal as their breeding stock for the future, rejecting the rest. This is called artificial selection, and it can produce astonishingly fast results.

After 10 000 years of farming, the human race now relies on about a dozen species of domesticated animals, and more than 100 species of domesticated plants. Some of them – usually minor sources of food – are practically identical to their relatives in the wild, because there has been little systematic effort to improve them. Others, including a whole range of plants and animals, from chickens to carrots, are still similar to their wild relatives, but they differ from them in size and shape. However, many of our chief sources of food no longer fall into either of these categories; instead they have been altered so much that they look quite different from their wild ancestors, and in some cases – including crops such as broad beans, lentils and date palms, and also horses and cattle – their wild ancestors no longer exist. These plants and animals have now been completely detached from the world of natural competition, and they rely on us for survival.

This is the remarkable outcome of the story that began over 2 million years ago, when early humans first learnt to shape materials around them. Then humans depended on what their surroundings could supply. Today we decide what our world will provide, and what other forms of life will share it with us.

ARTIFICIAL SELECTION *After many centuries of domestication, today's cattle look very different from their wild ancestors.*

POPULATIONS THAT GROW

In the two centuries since 1800 the human population has expanded at an unprecedented rate, eclipsing thousands of years of steady growth. This brief period has been a time of far-reaching technological achievement and social change.

By studying both fossils and human artifacts, palaeontologists and archaeologists have been able to reconstruct many aspects of life in former times. With a little imagination we can envisage how stone tools were chipped away, how fire came to be harnessed, how crops were raised, and how wild animals were eventually tamed. But one intangible feature of those distant days will probably always be beyond our grasp: how thinly the human race was spread across the globe.

Even at the dawn of agriculture, about 10 000 years ago, the human family was probably only 4 million strong. Some people still followed a nomadic lifestyle, wandering large distances in search of their food, while others had settled down to work the soil. But wherever and however they lived, people were hardly cramped. If everybody alive at that time had been evenly distributed across the inhabited world – without making any allowances for the terrain – each man, woman and child could have had about 13 sq miles (33.5 km²) of land to him or herself.

Since then something quite phenomenal has happened. During a tiny sliver of time, equal to just a few thousandths of our total history, we have emerged from relative obscurity to occupy a unique position in the living world. Instead of numbering a few million, we now total several billions, and we outweigh all other forms of animal life on land. This extraordinary success story has reshaped the world, and will have far-reaching effects as our numbers continue to grow.

THE STORY OF EGYPT

To find out what has catapulted humans into such prominence, demographers – people who study population – often turn to regions where the inhabitants

THE MILLING CROWD *At a fair in northern India traditional costumes create a brilliant spectacle, an accurate reflection of this densely populated land.*

have followed the same lifestyle for a very long time. One such place is the Nile Valley, where an ancient and self-contained culture has existed for more than 6000 years.

The fertile strip of land that borders the Nile is isolated by the vast emptiness of the Sahara Desert, which has helped to minimise contact with outsiders. Over the centuries trading caravans have snaked their way into the valley and across the delta, and invading armies have arrived by land and by sea, but the inhabitants of this landscape have experienced relatively few of the drastic upheavals seen in other parts of the world. For this reason the valley has been described as

an almost ideal laboratory for studying the ways human populations rise and fall.

Historians have studied surviving records and compiled population estimates for Egypt stretching back many millennia. In 4000 BC the people of the Nile lived by hunting and gathering, as well as by farming. The total population was probably about 350 000 people, which works out at about 28 per sq mile (11 per km²). It is not much by modern standards, but it is an unusually high figure for Neolithic times and reflects the fertile nature of this thread-like oasis, which was nourished by the rich silt laid down every year during the Nile's annual floods.

CRUCIBLE OF CHANGE
Riverside dwellers tend floating baskets of vegetation by the banks of the Nile in Sudan. For many centuries the Nile Valley's population changed only slowly. Its recent dramatic growth has been mirrored around the world.

As agriculture expanded and became more important, the population slowly grew. It reached about 1 million by 3000 BC, and 2.5 million by the year 1320 BC, during the reign of the Pharaoh Tutankhamun. This was a period of extraordinary cultural achievement, which saw the building of great temples and pyramids, but as far as human numbers were concerned it was a time of unremarkable growth. In good times the population could increase by an annual rate of up to 0.2 per cent, but in bad years it could easily slip back by the same amount.

By the 1st century AD, when Roman emperors had replaced the pharaohs, the Nile Valley was home to about 5 million people, a total that amounted to many times the pre-agricultural level. Farming had already transformed the valley floor, and awesome monuments provided ample evidence of a highly organised society. But from this moment onwards, population growth went into reverse. Political instability and poor harvests triggered a slow but steady decline, and numbers in the valley did not struggle back to the 5 million level until the 14th

FARMING ON THE NILE
Egyptian tomb paintings provide a remarkably detailed record of life in the Nile Valley long ago. Right: A surveyor uses a line to measure out a field. Below: A man uses a sickle to harvest grain. Both paintings are more than 3000 years old.

century, when the days of the pharaohs were already long past. At that point, when the population looked poised to continue its gradual climb, a hammer-blow fell. In 1370 bubonic plague broke out and spread rapidly along the valley and throughout the delta. The population spiralled downwards, and by 1600 it stood at just 2 million – back to where it was nearly 3000 years before.

This precipitous slump proved to be a turning point in the valley's human history. It was the steepest fall on record, but it also ushered in a new era of change. This time, when the Nile Valley's population began to recover, it did so in a truly spectacular way.

Initially the increase was slow, but by 1800 the population had climbed back to nearly 4 million, and within a further 50

years – just two generations – it was back to 5 million once more. However, it did not stop there. By 1897 the population had reached nearly 10 million, and within a further 30 years there were 14 million people along the Nile. By 1960 the population had almost doubled again, and by 1980 it stood at 42 million and was growing at a record 3 per cent a year. Something remarkable had happened in the Nile Valley – something that has occurred in many other parts of the world since the modern era began.

HAZARDS OF THE PAST

The last two centuries have been fascinating and probably unrepeatable times. After millennia of faltering growth the human race has suddenly expanded at a breathtaking rate. This brief period has been a time not only of booming population but also of remarkable scientific achievement and dizzying technological change. Instead of being confined to particular regions, our influence has spread out like a shock wave and has reached across the whole of the planet.

In some places, including the United States and most of Europe, this upward momentum now seems to have evaporated. Here the population has reached a larger but apparently stable level, or is increasing only slowly. In other regions, including Egypt and many countries in the developing world, the upswing is still in progress, although the rate of increase is generally beginning to fall. All across the world the pattern is the same: a long period of gradual increase, followed by a sudden and quite unprecedented change.

How has this remarkable leap in numbers come about? The answer is not that we reproduce faster than our distant forebears – in fact, the opposite is probably true. Instead, it has happened because we have become much better at surviving the hazards of everyday life. As a result, many more people live long enough to have children of their own.

Since the days of prehistory, humans have faced two natural hazards: disease and famine. The graver of these has always been famine. When people started to farm, long after humans had first evolved, life became more secure because crops could be stored, and because more food could be raised by

opening up new ground – if the labour was available. In Egypt, for example, farming began close to the Nile but slowly spread out towards the fringes of the river valley. But improvements like these developed steadily over thousands of years, without creating a population boom. The major impact on human numbers has been made not by farming but by the fight against disease.

In human history, disease has had an enormous impact. Malaria – thought to be the greatest killer of all – has been responsible for about half of the human deaths since the Stone Age, if wars and accidents are excluded, while a host of other infectious diseases, from diphtheria to smallpox, bubonic plague and cholera, have also taken their deadly toll. Together

these diseases have brought millions of lives to premature ends.

For most of the human span on Earth, people had no idea of how to escape these mysterious threats to their safety, apart from fleeing from an area of infection. When humans lived in small groups and followed a nomadic way of life, disease probably had a relatively modest effect on

WATER IN A RAINLESS LAND
Agriculture in Egypt depends on rain that falls far away in the highlands of Ethiopia. This reliable water supply has sustained farming for thousands of years.

population growth, but once they settled down and started to live side by side the scope for infectious diseases increased hugely. No one knew where these diseases came from or how they spread, but one thing was certain: they could strike suddenly and with deadly results.

LENGTHENING LIVES

In the 14th century AD the people of the Nile Valley were not alone in enduring the plague. The epidemic had already swept across Europe, and between 1347 and 1349 up to a third of the population died. In a world that depended on the labour of people and their animals, this sudden depopulation threw society into turmoil, and it also had some dramatic effects on the landscape. Villages were abandoned and fields left unploughed, allowing the natural vegetation to creep back and reclaim the ground that it had lost. In time most of the stricken settlements were restored, but some were so blighted that they never struggled back to life. Even today in some parts of Europe, grass-covered mounds and ridges serve as poignant memorials to these plague villages, where the disease had such catastrophic results.

Bubonic plague reached Europe from Asia, and arrived long before people understood how such diseases were transmitted. However, even in medieval times one feature of its

TRIUMPH OF DEATH *Although ghoulish to modern eyes, this Italian painting simply underlines the precariousness of life in medieval times.*

spread was obvious enough. People noticed that the plague seemed to travel with its victims, so one way to avoid it was to keep outsiders at bay.

In northern Italy, towns and cities began to adopt this measure and imposed a period of isolation on anyone wishing to enter their gates. The stipulated period of 40 days – known as a quarantine – was long enough for fortunate visitors to prove themselves healthy, and for the unlucky to sicken and die. Quarantines did not eradicate the plague, because the disease is actually spread by fleas and rats, rather than by plague sufferers themselves, but by keeping all outsiders out they did help some places to avoid infection. Quarantines were an important innovation because, in a world characterised largely by resignation in the face of disaster, they marked the beginning of practical measures to promote public health.

By the 18th century the idea of containing disease was well established, and instead of cities sealing themselves off from infection, things sometimes happened the other way around. In 1720, for example, a late outbreak of the plague flared up in the French port of Marseilles. To prevent the disease from spreading to surrounding

areas, the city was isolated by a *cordon sanitaire*, which was cleared of people. When the zone was accidentally breached, regiments of cavalry were brought in to establish a new zone farther back from the city. This second cordon worked. The plague was contained, and the outbreak eventually died out.

Other public health measures helped to cut deaths dramatically. The technique of

held down by the large number of infants and children who succumbed to disease. In the 1840s the city's water supply and drainage system were overhauled, and the danger from waterborne diseases was reduced. By 1870 average life expectancy had risen to 37, and by 1900 it stood at 50 – at a time when most of today's medicines were still unknown.

During the 20th century, with the help of antibiotics and modern surgery, life expectancy has risen even more sharply. In many developed countries men and women can now expect to live well beyond 70, and this increase shows no signs of coming to a halt. Compared with our ancestors of ten millennia ago, who had a life expectancy of less than 20 years, we have become extraordinarily long-lived.

WHEN GROWTH TAKES OFF

In nature the numbers of each species rise and fall according to local conditions, but usually stay within certain limits. However, just occasionally a species experiences the equivalent of a gambler's lucky break: its food may suddenly become abundant, its most important predators may be attacked by disease, or the weather may be just right for several years in a row. Whatever the reason, the outcome is the same – a period of rapid growth.

Improvements in public health have brought about one of these rare biological events in our own species, and the results have been spectacular. From a global population of just under 700 million in 1700, human numbers rose to about 1.6 billion in 1900, and are expected to exceed 6 billion early in the 21st century. Because the rate of growth has itself grown, the number of humans added each year has soared, so the time needed for the world's population to double has dropped from about 1000 years, when agriculture began, to about 40 years in the middle of the 20th century.

Taken to its mathematical extremes, this kind of growth produces mind-boggling results. For example, a single female mouse can theoretically produce more than 500 descendants in the course of a year, while a housefly – one of the fastest reproducers in

preventing smallpox by immunisation spread through Europe in the 19th century, and at the same time the importance of hygiene started to become appreciated. In one of the first examples of epidemiology – the scientific study of the way diseases spread – a British doctor called John Snow analysed a London cholera epidemic in 1854, and noticed that many of the cases were clustered in a handful of streets that shared a single water pump. The pump had been set up not far from a sewer, and when the resourceful doctor had its handle removed, so that people were forced to collect water elsewhere, the rate of infection plummeted.

These advances were made before anyone knew what caused infectious diseases, and they relied on nothing more advanced

FLAWED PROTECTION
This 18th-century cartoon shows a doctor during an outbreak of the plague. His smoke-filled hat was just one of the bizarre methods used to fend off the disease.

than cooperation and common sense. Their effect, however, was phenomenal. Together they triggered one of the greatest changes that our species – and the world – has ever seen. The figures speak for themselves. In 1840 the average life expectancy of a person living in the French city of Lyon was only about 32 years. Some of the city's inhabitants managed to live to what we would recognise as old age, but the average was

AN ISLAND OVERWHELMED

Famous for the enigmatic statues or *moai* that stare out to sea from its windswept slopes, Easter Island is a remarkable example of what can happen when a human population outstrips its resources. This remote fragment of land in the South Pacific was uninhabited until around AD 400, when it is thought that a group of about 50 Polynesian settlers made a chance landfall on its shores. At that time the island was covered in forest, but the new arrivals soon began to clear the trees and grow food on the fertile volcanic soil.

Against all odds, the settlers flourished. Over the next 1000 years this most isolated of all human populations rose to between 5000 and 20 000 people – estimates vary – crammed onto an island measuring only 64 sq miles (165 km²). Their water came from Rano Kau, a flooded crater in the island's south-east tip, and they lived on crops such as taro, sweet potatoes and bananas, on fish, and also on chickens, which were brought to the island by the original settlers. Metal was unknown, and the islanders worked the land with wooden implements. They cut trees with knives made of obsidian, a black volcanic glass.

For the first 1000 years the Easter Islanders seem to have prospered, despite growing tensions between rival clans. The cult of the *moai*, which involved carving 15 ton statues, and then shifting them several miles from quarries to the sites where they were set up, shows that the islanders had time for much more than simple survival. But little by little things began to go wrong. As the population grew, farmland had to be divided into smaller and smaller patches. More and more trees were felled, and as the forest disappeared it became increasingly difficult to conserve the soil, to build or repair canoes, or to find the fuel needed for cooking.

When the first Europeans set foot on Easter Island in 1722, they found a society that was teetering on the edge of implosion. Over the following 150 years food production fell, the *moai* were toppled during a series of bloody wars, and the population underwent a precipitous decline.

During the 19th century the last of the island's trees was cut down, and the surviving islanders were menaced by raiding parties from the South American mainland.

Today Easter Island is a place of windswept grass. The patchwork of gardens that sustained the original islanders has gone, leaving only the haunting *moai* as reminders of former times, and of a people that ran out of space.

LEGACY FROM THE PAST
Easter Island's statues were created by the work of many hands. After the island society collapsed, the labour pool vanished, and only the statues now remain.

the animal kingdom, with a life cycle lasting as little as 20 days – could soon swamp the entire planet with its buzzing progeny. Even an elephant, giving birth at the leisurely rate of one calf every four years, could leave 100 descendants during its 60 year span. However, in nature this kind of explosive growth always has the same outcome. After a while space or food begins to run out, and before many generations have passed the growth in numbers comes to a halt. The levelling off is often followed by a population slump, as numbers fall back to more normal levels.

TOWARDS A STABLE STATE

This natural cycle of boom and bust might seem like an ominous sign for our own species. Indeed, one of the earliest demographers, the English clergyman Thomas Malthus, warned that human numbers would inevitably expand like those of other forms of life until wars, famine and disease kept them in check. But in the two centuries since Malthus put forward his ideas, things have turned out to be less straightforward than he imagined. The human population is now six times the size it was in 1798, when he first published his ideas, but, despite this

THE WORLD'S LARGEST POPULATIONS

	Population (millions, 1994)	Projected population (millions, 2010)		Population (millions, 1994)	Projected population (millions, 2010)
China	1191	1388	Mexico	92	118
India	914	1189	Germany	81	80
United States	261	297	Vietnam	73	98
Indonesia	190	240	Philippines	66	88
Brazil	159	199	Iran	66	95
Russia	148	143	Turkey	61	78
Pakistan	126	210	Thailand	59	67
Japan	125	127	United Kingdom	58	60
Bangladesh	118	163	France	58	60
Nigeria	108	168	Egypt	58	81

huge increase, improvements in health, agriculture and technology have managed to keep pace with growing numbers and have allowed the human species to thrive.

While this boom has been taking place, some less conspicuous changes have been under way. In many parts of the world, people not only live much longer than they once did, they also have smaller families. Where this has happened the growth rate has begun to drop, and in a few countries it has even come to a halt. The result of this kind of change is a jump from one roughly stable population level to another, a phenomenon known as a demographic transition.

Different regions of the world are experiencing this transition at different times, and at different speeds. In Sweden the transition began in the early 19th century and ended in the 20th, and it produced a population that is roughly four times its former size. In Mexico

DISINFECTION POINT *An illustration from 1911 shows the use of a disinfectant spray to prevent the spread of cholera. Public health measures have reduced the impact of many diseases.*

it began in the 20th century and is expected to end in the 21st, producing a population perhaps ten times its previous level. In some African countries it has only recently begun, and here it may create the largest growth of all, with populations 20 times bigger than they once were. But despite these regional differences, a similar pattern seems to be at work throughout the world: like a car accelerating and changing gear, the numbers jump upwards and then eventually level off.

This leaves us at a unique point in our species' history. During the 20th century the planet's human population has reached a size that dwarfs that of earlier times, but the growth that fuelled that change is starting to run out of steam. After almost 200 years of rapid increase, the end of an extraordinary era is in sight, and the world's population is unlikely ever to double again.

According to recent research, the global population is likely to peak at around 10 or 11 billion in the year 2075. The Earth will certainly be more crowded than it is today, but whether it will be overloaded is a harder question to answer. With the help of tools, agriculture and modern technology we have already broken through the barriers that hold back other forms of life, and have transformed the world in which we live. On a more populous planet, where there are no new frontiers or untouched lands, ingenuity will be even more important for our survival.

LOCAL CHANGE

2

WHEAT LINES *Cereals are grown on an industrial scale in regions like Victoria, Australia.*

FOR MOST PEOPLE, THE LIFESTYLE FOLLOWED BY OUR ANCESTORS IS AS REMOTE AS IT IS UNFAMILIAR. INSTEAD OF WANDERING IN SEARCH OF FOOD, WE LIVE IN PERMANENT HOMES WHERE ALL THE NECESSITIES OF LIFE ARE WITHIN EASY REACH. THIS NEW KIND OF LIVING HAS TRANSFORMED SOCIETIES AND HAD DRAMATIC EFFECTS ON OUR SURROUNDINGS AS WE MAKE INCREASING DEMANDS ON RESOURCES AND SPACE. TEN THOUSAND YEARS AFTER THE SETTLED LIFE BEGAN, THE RESULT IS A WORLD REWORKED: EXPANDING CITIES HOUSE A GROWING POPULATION; INTENSIVELY WORKED FARMLAND PROVIDES US WITH FOOD; AND ENGINEERING ACHIEVEMENTS, FROM MINES TO DAMS, SUPPLY US WITH RESOURCES MOST OF US NOW TAKE FOR GRANTED.

EERIE GLOW *An oil refinery in Nova Scotia lights the night sky.*

RESOURCES FOR LIFE

Like all living things, we need energy and water – the basic resources that sustain life. The way we obtain those resources, and the way we dispose of the waste we create, has major implications for the world about us.

Shortly after dawn on a summer's day, a bumblebee climbs sluggishly from its underground home. It waits for a few minutes while its body temperature gradually rises, and then takes off on rapidly whirring wings. Some time later, its departure is followed by one of a different kind. A human being emerges from a house and climbs into a car. The door clicks shut and the engine starts, and the car is soon making its way through the traffic towards the other side of town.

The clock now advances to midday. As it has done about every 20 minutes since the day began, the bee returns to its nest, laden with pollen and sugar-rich nectar. By chance, this is also the moment at which the driver comes home. After retrieving several bags from inside the car, the heavily laden figure disappears indoors. Only a few hours have passed, but during that time two very different forms of life have fulfilled the same urgent imperative: the need to find food.

Making comparisons between humans and animals is always hazardous, and comparing ourselves with creatures as simple as bumblebees risks

ENERGY ON TAP *Producing power inevitably involves waste heat. Here billowing clouds of steam escape from the cooling towers of a power station.*

CREATING FIRE *By rubbing specially shaped fire sticks, a Kung man from Namibia soon generates the heat needed to produce a flame.*

incredulity. But despite our unique position in the living world we still share many characteristics with less sophisticated forms of animal life. We need food, water, warmth and shelter, and we get these just like all other consumers in the natural world – from bees to chimpanzees – by using what our surroundings have to offer.

Animal life has existed on Earth for many millions of years, so consumers are nothing new. But the way humans, particularly modern humans, consume resources is quite different from anything seen before. Instead of remaining constant, our everyday needs keep growing and, even more importantly, they change. They are so complex that they affect both our immediate surroundings and places far beyond.

ENERGY: THE ULTIMATE RESOURCE

For all forms of life, energy is the key to existence. It drives the chemical processes of living cells and generates warmth and movement. Animals get their energy from food, and release it by breaking the food down into its chemical components. The energy they release is nearly always somatic, meaning that it is channelled through the body itself, and not through anything outside.

Even a simple animal such as a bumblebee has a complex energy budget, which ensures that the energy it uses is allocated in the most efficient way. For example, when a bumblebee emerges from its nest, it shivers its wing muscles to warm them up, and uses some of its energy in the process. However, most other parts of its body do not need to be at such a high temperature, so to save energy the bee thriftily leaves them cold. The bee's energy budget also shapes its behaviour in more far-reaching

FUEL FROM FLOWERS *Probing deep into a clover flower, a bumblebee is rewarded with a droplet of nectar – a high-energy fuel.*

ways. It dictates how far it will go to find food, how often it will return home, and even when it will give up foraging altogether and hibernate instead. Energy – or the lack of it – is the resource that shapes the whole of the bee's life.

Animals often store energy in the form of food, and dip into their stores when times are hard. But no matter how much food an animal has stored away, it can release energy only at a certain rate. A bee's energy output, for example, has a built-in maximum of one bee-power. If its cells broke down food any faster, they would be killed by the heat they generate.

A human being's power output is vastly greater than a bee's, but the underlying principle still holds true. However, as early humans discovered – no doubt to their excitement and alarm – there is no upper limit to the energy that can be released by non-biological means: the more fuel is thrown on a fire, the more energy is released. Once

humans had learned to control fire they could, for the first time, begin to tailor their energy output to match their lifestyle, rather than the other way around.

According to today's calculations – which can only ever be approximate – the energy that early humans obtained from campfires may have been about equivalent to the energy they got from their food. It was not much by modern standards, but it pointed to the way ahead. Once agriculture had become well established, energy use climbed again. Early farmers probably obtained at least five times as much energy from wind, water and animals as they did from the things that they ate. But the real change did not come until much more recently, with the widespread use of fossil fuels. Today the average person in the developed world uses about 60 times as much energy as our hunting ancestors. Much of this is used in factories, offices and transport, but a significant amount is put to work at home.

ENERGY AND THE EARTH

Until the early 19th century, fuel was a very visible part of everyday life. Whether it was wood, coal or candle wax, it had to be bought, brought home and then burnt. Gathering these fuels had already had consequences for local landscapes. In Europe, woodlands were often managed for timber and firewood, and itinerant charcoal-burners produced fuel for smelting metals. Where the climate was damp and cool, peat was dug out of bogs, creating a patchwork of pits in the spongy ground.

Sometimes the quest for fuel had far wider effects. Candle wax, to take another example, was originally not the wax we know now, but tallow – a mixture of greasy fats that burned with a smoky flame and a

pungent smell. These fats once came from domestic animals, but from about 1750 onwards a superior and much cleaner-burning alternative was discovered in the form of spermaceti, a waxy substance in the heads of sperm whales. Spermaceti candles, together with whale-oil lamps, soon became popular, and sperm whales were widely hunted in the North Atlantic. However, catches quickly fell because of overhunting, and to make up for the shortfall whaling ships fanned out into the Indian Ocean and the Pacific. By 1820 they were operating off the coast of Japan, driven thousands of miles from their home ports by the simple quest for domestic light.

BRIGHTER LIGHT · *The invention of gas mantles in 1885 increased the light output of gas lamps by about six times, and lit up homes and streets.*

POWER FROM COAL · *Wreathed in smoke and steam, early coal-burning power stations generated the electricity needed by growing towns and cities.*

Had this trade continued, sperm whales might well have become extinct. But during the 19th century two new fuels arrived which were to have an even greater impact on the natural world. One was coal gas, which appeared early in the century, and the other was mineral oil, or petroleum. This subterranean oil had been known for centuries, but its use as a fuel began only in the 1850s when it was discovered that it could be distilled to produce kerosene. When Edwin Drake, an American railway conductor, sank the world's first oil well in Pennsylvania in August 1859, the new fuel became available on a commercial scale.

WEB OF WIRES *High-voltage power lines, seen here crossing Austrian farmland, are one of the most visible signs of our increasing use of energy.*

The use of oil has had immense repercussions on our planet, but in one way piped coal gas had its own unique significance. When the first gas pipelines were laid and the new gas lamps flared up in fashionable drawing rooms and along city streets, the era of instantly accessible energy began. For the first time, energy was 'on tap'. People whose homes were connected up to the supply could use as much gas as they could afford, and did not have to concern themselves with where the fuel came from or how much there was left.

When gas was superseded by electricity, from about 1880 onwards, instantly accessible energy spread even further. Power lines

FARMING THE SUN

Solar power currently supplies only a tiny fraction of the world's energy needs, but different ways of gathering this inexhaustible source of energy may soon transform the world's landscapes. There are two principal ways of harnessing solar energy: in direct methods the energy in sunlight is used to generate electricity, while in indirect methods it is used to produce fuels which release energy when they are burnt.

To generate electricity, sunlight must first be collected and its heat accumulated. In some solar energy stations this is carried out by rows of trough-like reflectors that focus the light onto parallel pipes; in others the light is focused by parabolic reflectors, or arrays of flat mirrors angled so that the light strikes a central boiler. The boiler produces steam, and this drives the turbine that generates a current. The trouble with these systems is that they involve a number of steps and

need to produce high temperatures, both of which reduce their efficiency.

Another way of generating electricity is known as photovoltaics. This produces electric energy by using silicon semiconductors, or solar cells, and does away with intervening steps and high temperatures. Using this kind of technology, a solar power station at Carrisa Plain in California currently generates about 7 megawatts of electricity (compared with several thousand megawatts generated by many hydroelectric stations), and a 100-megawatt station, also using solar cells, is being built in the Nevada Desert. Solar cells are expensive to make, and harness less than 10 per cent of the light energy that falls on

POWER WITHOUT POLLUTION
In California's Carrisa Plain power plant, computer-controlled photovoltaic panels harness the energy of the Sun.

them, but many energy experts believe they have a promising future.

In the second method of harnessing solar energy, sunlight is used to grow plants that produce fuel. The fuel is either a plant oil or alcohol produced

by fermenting plant sugars. Growing fuel is less efficient than generating electricity, but it has the important advantages that it can be carried out in a range of climates, and the energy it produces is in a portable form.

fanned out from cities and towns to reach rural areas beyond, and the web of energy production and use became ever larger and more elaborate.

ENDLESS LIGHT

Every year the Earth receives about 15 000 times more energy from the Sun than the human race consumes. Seen in this perspective, the extra fifteen-thousandth contributed by our activities sounds quite trivial. But because of the way we collect and use this energy, our fifteen-thousandth has had some remarkable effects.

One of these is visible when our ultimate power source – the Sun – sinks below the horizon, and night sets in. At one time night was a period of more or less total darkness, relieved by the light of the Moon or the faint flare of the stars. Like many animals, people used the Moon as a lamp and timed their activities to coincide with its phases. In England some of the world's earliest industrialists formed a group called the 'Lunar

NIGHT LIGHTS *In just over a century, electric light has transformed city life. In this spectacular panorama the lights of Hong Kong gleam and sparkle in the night.*

Society'. The lunar connection was a purely practical one – the full Moon helped members to make their way home.

That world of darkness has given way to a very different one today. Thanks to readily

available energy, night – described in the Bible as a time 'when no man can work' – has been transformed by light that floods into the sky from cities, homes and street-lamps. This light often masks the light of the stars above, and is clearly visible from space. But among these earthly sources of light a few seem out of place. Instead of shining from centres of population, some gleam in isolation from the otherwise featureless voids of deserts or unlit seas. These beacons are gas flares – the nocturnal evidence of

AN EXPLOSIVE DISCOVERY

The ancient Chinese used natural gas over 3000 years ago, but its potential lay largely unexploited until 1821, when a water well in Fredonia, New York, started to produce bubbles from deep below ground. The bubbles contained natural gas, and this accidentally caught fire, creating a giant flare. Following this incident, further wells were dug in the area, and the gas was piped to a local hotel. From this modest beginning, natural gas gradually became an important domestic fuel.

one of mankind's greatest enterprises in the search for energy, the endless quest for oil.

Since Edwin Drake opened his first well, oil's versatility has given it an unrivalled importance in our use of energy. Whereas Drake produced about 35 barrels a day, pumping it up from a depth of 70 ft (21 m), the world's daily production now stands at about 2.5 million times that amount. Some still comes from deposits near the surface, but the depth of wells has steadily increased, and a few reach 10 000 ft (3050 m). A complex technology has developed which allows engineers to draw up profiles of the Earth's crust in an effort to identify places where oil is likely to be found. That technology is improving all the time, but exploration is still a gamble: overall, only one in ten test wells produces oil in viable quantities.

Our demand for oil has taken engineers and surveyors to places as far apart as tropical rain forest and polar ice, opening up areas unaccustomed to human presence. When oil is discovered in worthwhile quantities, it has to be extracted, carried away and refined, and at each step in this process there is a risk of unplanned and sometimes dangerous consequences, from wellhead explosions to oil spills at sea.

THE EFFECTS OF ENERGY USE
Energy has to be turned from one form into another to be used. Putting a match to a candle, for example, converts chemical energy into light and heat, while flicking a switch achieves the same end from an electric

current. An important difference between these two ways of lighting a room is that the candle actually breaks down a fuel. An electric light bulb works in a different way. Electricity delivers the energy it needs, but the process of producing that energy is carried out somewhere else.

Because many homes are now powered largely by electricity, the consequences of using energy can be easy to overlook. Unlike a candle, a light bulb does not burn with a smoky flame, and refrigerators, televisions and computers do not fill our homes with the products of combustion. But for most homes the ultimate source of energy is still combustion, and this combustion is now carried out on the grandest of scales.

Today's fuel-burning power stations have an immense energy output, and their impact on their surroundings is of similar stature.

FROM THE DEPTHS *Perched above Europe's continental shelf, a North Sea oil platform burns off waste gas.*

CLEAN POWER *A hydroelectric plant generates power in a valley of the French Alps, using water from a nearby reservoir. Once built, plants like these produce little pollution.*

One of the problems in generating electricity is that it can never be 100 per cent efficient, and inevitably involves generating waste heat, disposed of by, for example, evaporating water in cooling towers. However, heat pollution is a minor problem; burning fuels have much more far-reaching effects.

When electricity is generated by burning coal, oil or gas, compounds of sulphur and nitrogen find their way into the air. These are among the causes of acid rain, which can kill trees and damage life in lakes. Some of these substances can be removed at power stations by chemical means – although the technology has yet to be widely adopted – but there is a greater problem that cannot be tackled in this way. When they are burnt, all fossil fuels add carbon dioxide to the air, and this contributes to global warming by making the Earth's atmosphere trap more of the Sun's heat.

For some energy experts, this problem provides a powerful reason for making energy without using fossil fuels. At present the alternatives contrast sharply. On the one hand is nuclear energy, the energy that lies within matter itself, and which binds atomic nuclei together; on the other is the self-renewing energy of the Sun, which bathes the Earth in its heat and provides the driving force behind flowing water and moving air.

Nuclear power currently supplies about 5 per cent of the world's energy, while hydro-electricity – by far the most important source of 'renewable' energy – supplies about 6 per cent. Despite its initial promise, nuclear power has turned out to be unexpectedly expensive, largely because of the precautions needed to contain its dangerous fuel. At some point in the future, power may be generated more safely and cheaply by combining atoms instead of splitting them, but at present a commercial version of this process – called nuclear fusion – is still a long way off.

After a succession of accidents, few people need to be

MYSTERIES OF THE PAST

In the flat landscape of Norfolk in eastern England, about two dozen freshwater lakes are scattered close to the coast. Known as the Norfolk Broads, they long puzzled geologists. Originally they were thought to have been formed by the build-up of silt and changing sea levels, but now a very different explanation has come to be accepted.

Instead of being made by nature, the lakes are almost certainly the result of human activity. Although disguised by the passage of time, straight edges, sharp corners and steep sides show that they were deliberately excavated long ago. Their exact age is uncertain, but the digging continued for generations and was probably well advanced 700 years ago. The object of these excavations was peat. Produced by the waterlogged remains of plants, this fibrous fuel was cut out in blocks and used to provide heat for homes. It was also used for warming pans of seawater to obtain its valuable salt.

Altogether, more than 850 million cu ft (24 million m³) of peat were removed from the ground. Peat-digging came to a halt when the sea level rose and the peat pits were flooded.

DISGUISED BY TIME *A lone windmill punctuates the flat landscape among the Norfolk Broads. The flooded peat diggings are a magnet for freshwater birds.*

reminded of the hazards that nuclear power brings, but renewable energy also has drawbacks. Hydroelectric dams often drown valuable land and disrupt natural ecosystems, and take many years to recoup their costs. Wind farms create visual pollution and generate only modest amounts of power, while converting sunlight into electricity also works on a limited scale and only in certain parts of the world. Improved technology will undoubtedly increase the viability of renewable energy, but fossil fuels will continue to play a large part in life for many years to come. To minimise their effects, many energy scientists advocate a different course of action – reducing the amount of energy that we actually use.

We are still beginners in the art of energy economy. Some birds use air currents so they can fly with almost no energy expenditure at all. Fireflies change chemical energy into light but produce almost no waste heat. Compared with these living machines, our machines and devices lag far behind.

However, the history of artificial light – to trace just one form of energy use – does show that we are at least heading in the right direction. Oil lamps and candles are very inefficient, because they release far more energy as heat than as light. The first gas lamps were even worse, and the first electric lights also released most of their energy as heat. Even standard light bulbs, with a tightly wound tungsten filament, are only about 7 per cent efficient, which means that 93 per cent of their energy is wasted. Today's low-energy fluorescent light bulbs are far better than this, and deliver at least eight times as

HARNESSING THE WIND *Like posts in a straggling fence, wind turbines crown the hilltops of California's Altamont Pass.*

much light for each unit of energy, and future forms will undoubtedly be more efficient still. With cleaner sources of power, better designed machines, and insulating materials that rival those found in nature, there is a real chance that our energy impact could be steadily reduced.

A LIQUID DIVERSION

In the span of human technology, energy at the flick of a switch is still a recent development. By comparison, water at the turn of a tap – or at least at the unplugging of a pipe – has a much longer history. At the height

of the Roman Empire, about 2000 years ago, aqueducts carried water into major cities so that it could be distributed, and pipes made of lead or wood delivered running water direct to homes.

Today's water supply systems are far more sophisticated, but so too is our thirst for this essential resource. Now that water is so accessible, our demand for it has grown, much like our demand for energy. In 1996 a group of American scientists tried to estimate this demand by comparing it with the world's annual runoff, which is the water that flows away over the land's surface. Excluding floods and runoff in remote areas, they found that humans now use more than

half the fresh water that is available, and that the figure is rising fast.

To stay alive, most adults need between 3 and 4 pints (1.7 and 2.3 litres) of water a day. This is about average for animals of our size, although it rises sharply if we start to overheat. But just as with energy, the amount of water we need to maintain our bodies is only a small fraction of the amount we actually use from day to day. At home we use water not only for drinking but also for washing, for cooking and for carrying away waste; we use it, too, on cars and in gardens. In a typical European city, this alone brings the daily consumption of water per person to about 70 gallons

WATER SUPPLY *Built some 2000 years ago, the Pont du Gard in southern France is a testament to the growing demand for water in Roman times.*

(320 litres), and in North America and some other parts of the world it is double this figure. Added to this is the water that is used on our behalf, which does not actually reach our homes. Irrigation accounts for about four-fifths of human water use, while industry also takes its share. Altogether, humans draw off more than 1400 cu miles (6000 km^3) of water from surface sources per year.

Statistics like these can be startling, but on their own they only tell one side of the story. This is because, unlike fossil fuels, water is rarely obliterated simply by being used. It does disappear if it is mixed into concrete, or if it is chemically combined in other ways, but in most cases it eventually returns – albeit in a dirty form – to nature's system of circulation. However, during its journey through our reservoirs, pumps, pipes and taps, water is diverted from the paths that it would normally follow. If the amount being diverted is small it has very little effect, but when it becomes very large the impact can be spectacular indeed.

Few places in the world demonstrate this more arrestingly than a small area of north-west Mexico, where the valley of the River Colorado broadens out before its meeting with the sea. The Colorado drains an area the size of France stretching across seven American states, and in an average year this basin captures the equivalent of 1 ft (30 cm) of water covering 15 million acres (6 million ha). Much of this falls as snow over high plateaus and mountains, and when the warmth of spring arrives, and the snow melts, the water begins a descent of up to 14 000 ft (4300 m) as it drops towards the distant coast.

That, at least, is what used to happen. Today, near the river's mouth in the Gulf of California, the only sign of this water is a snaking stream that slowly peters out in the desert sand. The remainder of the river's water is siphoned or pumped away before it reaches this point, creating a man-made water system that leads to a host of different destinations, from fields and golf courses to distant city homes.

The effect on the river's wildlife – particularly near the Colorado's estuary – has been dramatic, as wetlands have shrunk and the age-old cycle of drought and flood has been brought to a halt. But the effect on human life is also increasingly acute. This is because no matter how many dams are built and reservoirs are filled, every river can only have a finite flow, and once that flow is fully exploited no more water is available. In the cities of the Roman Empire, disputes over limited supplies of water were a part of everyday life, and in this arid part of North America, water shortages are having a similar effect.

DRAINING THE GROUND

When rain falls on the ground, about one-third flows away downhill. About another third evaporates, either from the ground itself or through the leaves of plants, while the final third trickles into the ground and begins a long subterranean voyage back to the sea.

Known as ground water, this underground resource has been exploited for millennia, ever since people first scraped away the soil and broke through the water table to the saturated ground that lay beneath. It was from these beginnings that the art of sinking wells developed, so that ground water could be reached even where it lay many feet below the surface. In most places ground water has to be hauled or pumped upwards, but in some it gushes out of its own accord because it is trapped under pressure between layers of impermeable rock. These artesian wells were widely dug in north-east France in the 18th century, and it is from the province of Artois that they get their name.

Compared with the water in a river, which seems to attach some urgency to its appointment with the sea, ground water is quite unhurried. As it seeps through porous rock from the soil above, its progress often amounts to just a few feet every year. In many places this water has been in transit more than a century, and in some – for example, Arizona and Israel – it is over 10 000 years old. In the Great Artesian Basin of Australia the ground water is thought to have fallen as rain up to 100 000 years ago. This ancient water is a liquid

WATER BENEATH THE SURFACE
Wells work by reaching down beneath the water table. Removing too much water can make a well run dry.

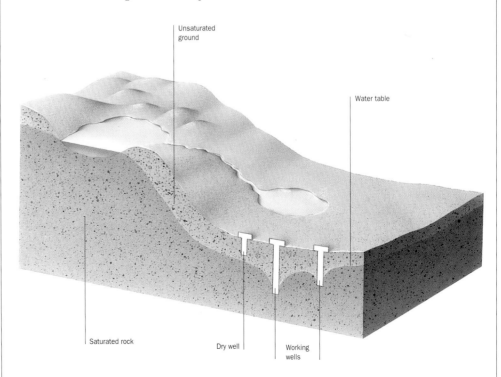

Unsaturated ground

Water table

Saturated rock

Dry well

Working wells

relic of times when the climates of these places were far wetter than they are today.

Given the vast volume of water which lies hidden underground – over 30 times the amount of fresh water on the surface – wells and boreholes might seem like mere pinpricks through the Earth's skin. But, because ground water flows so slowly, these pinpricks can have quite rapid effects. When water is drawn off the water table surrounding a well begins to drop, and the result is a 'cone of depression' rather like that made by a straw sucking up a very thick drink. If the water removal continues for long enough, the point of the cone eventually falls level with the bottom of the well. Either the well has to be dug deeper, or its working life comes to an end.

This is exactly what has happened in many large cities, such as London and Chicago. In London the number of wells sunk each year increased tenfold between 1860 and 1910, and the water table dropped by as much as 200 ft (60 m), while in parts of Chicago the drop has been even greater. No major city has yet run dry, but in areas where agriculture depends on ground water – such as the American Great Plains – the falling water table carries exactly this threat.

For cities that stand on sand or clay, falling water tables can have a much more visible outcome than empty wells. In Venice some of the world's finest buildings are slowly descending into the waves as the ground sinks beneath them. The reason for this is that, over many centuries, the water table has dropped and the ground has shrunk, carrying Venice with it. The Venetian authorities have reduced the rate of sinking by rationing water use and pumping water back underground, but whether this will save the city remains to be seen.

THE WORLD OF WASTE

For the world's natural consumers, water and energy from food are the essentials of life. But animal needs can also include materials of a quite different type. Bumblebees, for example, collect tiny strips of grass or pieces of leaf, birds gather twigs and mud, and beavers gnaw through saplings and ferry them downstream. Guided entirely by instinct, they fashion these materials into their nests and homes.

All forms of animal life work in these fixed ways. Each builder sticks rigidly to inherited designs and rarely comes up with innovations of any kind. Modern humans, however, are different. Instead of relying on inherited patterns of behaviour, we are constantly devising new uses for what the Earth can supply to us. Some of the things we create make life easier, safer or more comfortable, while others help us to learn, to communicate or to work. Others serve a uniquely human purpose: they simply make life more fun. But most of these products also do something else – they generate large amounts of waste.

As planes come in to land at Tokyo's Haneda airport, they often make their descent across Tokyo Bay. The coast follows a succession of sweeping curves, but the boundary between land and sea is masked by a flotilla of islands just offshore. Most of the islands – including the one on which the airport is built – have sharp corners and ruler-straight sides, showing that humans, rather than the forces of nature, have brought them into being.

These islands are Tokyo's answer to a particularly pressing problem: how to dispose of the city's daily output of 12 000 tons of rubbish. Some of the waste is burnt, but the rest is shipped out into the bay and piled up into man-made land. Each island is like a sandwich, consisting of alternating layers of soil and garbage. Some are used to berth ships or to provide stepping-stones for urban highways, and one even has a golf course.

Because they are scattered and made only partly of waste, these artificial islands fall somewhat short of making the record books. That dubious honour falls to a landfill site in Staten Island, New York, which is the world's largest single dump of 'pure' garbage. It covers an area of 3000 acres (1200 ha), holds about 2.5 billion cu ft (70 million m³) of waste, and grows bigger by about 100 000 tons a week. Rubbish on this scale has the capacity to make some major changes to the local scenery. The Staten Island landfill is spread over the low-lying ground of a coastal salt marsh, and instead of filling a hole it forms a flat-topped hill visible from several miles away.

For most of human history, people consumed little other than food, and they had little to throw away. Change was slow, so until something broke or wore out there was no point in replacing it with anything else. But with the onward march of technology, this low-waste lifestyle has almost completely disappeared. The kind of refuse piled up in Tokyo Bay or on Staten Island

NATURE'S ENGINEERS
Humans are not alone in damming and diverting water, but unlike beavers we abstract it on a massive scale.

consists of a complex mix of objects made of many different materials. The bulk is composed of packaging, paper and other kinds of everyday waste, but also included are things that have simply become obsolete or unfashionable. All this helps to form discarded weight of up to 4 lb (1.8 kg) per person per day.

With today's increasing emphasis on recycling and avoiding waste, it is likely that our output of waste will eventually start to fall. However, it will not happen before these giants have ample opportunity to swell much further. Geologists have a good understanding of the processes that affect natural landforms, but nobody yet knows precisely how these man-made piles will change during the passage of time. By investigating the interiors of large dumps, researchers dubbed 'garbologists' hope to find out.

INTO THE UNKNOWN

One fact beyond doubt is that these dumps are not simply inert. Their organic material – which includes food and garden waste – gradually rots down, releasing energy that warms the heap and in some cases even sets it alight. As well as generating heat, the decomposition process also produces methane, a colourless and flammable gas that is formed when matter decays in the absence of oxygen. If the methane comes into contact with burning waste, the results are potentially explosive. But as the gas moves up, other things move down. Rain percolating through the heap leaches out anything fluid that it meets, and this can include not only residues from organic waste but also dozens of different household chemicals from car oil to paint and bleach.

As time goes by the organic content of the heap declines, although not quite as fast as experts once supposed. In the airless interior of large heaps, decomposition takes place very slowly, so discarded bread or vegetable peelings can be intact – although flattened – more than a decade later. Paper, which is also made of organic matter, is more resistant to decay and so takes even longer to decompose, although it too will eventually disintegrate as bacteria set to work. This leaves the inorganic part of the waste: substances such as glass, ceramics, metals, plastics and bone, which bacteria are unable to digest. On average it makes up about two-fifths of the whole, and its potential life span is as great as the ground on which it stands.

The world's oldest surviving rubbish tips date back thousands of years, and were left

PROBLEMATIC PLASTIC Much domestic waste, as on this tip in Germany, consists of materials that did not exist a century ago. We are still learning how to deal with them.

by people who gathered food along the seashore. Their shell middens still give archaeologists some idea of what they ate, at a time when the Earth's resources were only lightly touched by human hands. Our gigantic rubbish heaps will tell a very different story to later generations. It is one that may mark our time as a uniquely profligate moment in human history, before the art of safeguarding resources achieved the prominence that it fully deserves.

FROM VILLAGE TO METROPOLIS

Almost half the population of the world now lives in cities.

Like giant and highly complex organisms, cities absorb

both people and commodities from far afield, and as they

grow they transform the land on which they stand.

If tradition is to be believed, the year 1626 saw one of the most important transactions in human history. For goods worth about $24, a Dutchman named Peter Minuit bought the entire island of Manhattan from its native inhabitants and thereby hastened a process that would transform the island itself and the land for miles around.

Today Minuit would not recognise Manhattan. It is still framed by tidal waters to the west and east, and is still separated from the mainland by the narrow Harlem River. But from Battery Park in the south to the Henry Hudson Bridge in the north, all traces of the original Manhattan island have vanished. Instead of densely packed conifers, skyscrapers now crowd Manhattan's interior, and concrete wharves stand along a shoreline that was once formed by rocks and mud. Every morning a human flood surges onto the island across bridges and through tunnels, and every evening the flow is reversed.

Few parts of the world have seen such an overwhelming re-drawing of a natural landscape, with such spectacular results. But extraordinary though it is, Manhattan's experience is only part of a process that has swept through a much wider area. Swollen by generations of immigrants, the population of New York has expanded across much of Long Island and the adjoining coast, spreading over forests, meadows and salt flats. Nearby cities have also grown, and as time has gone by, many of them have coalesced with their neighbours. This fusion has continued at an ever-increasing rate along much of America's east coast, and the result today is a string of supercities stretching over 500 miles (800 km).

Cities on this scale are artificial worlds. They are so big that they change the ground on which they stand, and they also have a profound influence over the environment around them. Some of these effects have been apparent for centuries, but others are only now beginning to be understood.

THE HISTORY OF URBANISATION

The first cities appeared about 5500 years ago, but it is only in the last 300 years that they began to assume the importance that they have today. Maps of the urban world in the year 1700 show that there were just five cities with populations of over half a million, and all were in the Northern Hemisphere. London and Paris were the two great cities of the west, while Beijing and Tokyo were their counterparts in the east. The fifth and largest of the cities was Constantinople (modern Istanbul), which had a population of about 700 000. Standing where Europe and Asia meet, it was the capital of the Ottoman Empire, which once stretched from the Balkans to southern Arabia.

A century later the number of major cities had risen to six: Guangzhou (Canton) had joined the ranks of the giants. Though still few in number, big cities were getting bigger; one of them, Beijing, had crossed a crucial threshold, becoming perhaps the first settlement in human history – with the exception of ancient Rome – to embrace more than a million people. However, for

HEART OF EMPIRE *Drawn in 1534, this map shows Constantinople, one of the largest cities of its time. The growth of New York (opposite) is one of the most dramatic examples of urbanisation. Less than four centuries ago, this scene was forest.*

most of the human race, life was still firmly anchored to farming and the land. In the rural world, knowledge of cities came mainly from travellers' tales, and these glowing reports often contrasted with a shabby reality.

By the year 1900 the Industrial Revolution had reshaped the world of work, and the number of great cities had increased to over 40. Sixteen of them had over a million inhabitants, and four of these had sprung up in North America, rapidly overtaking long established cities in other parts of the world. Rio de Janeiro and Buenos Aires became the first major cities in the Southern Hemisphere, while, in Australia, Melbourne and Sydney were also growing fast. Cities were sucking in people from the surrounding countryside, and also from overseas. As they expanded, the impact of such concentrated masses of people became very evident, and sometimes impossible to overlook.

Concern about the environment is often thought of as a 20th-century phenomenon, but historical records show that its origins – for city-dwellers in particular – are much earlier than this. Until the late 19th century most large cities had no sewage treatment of any kind, and liquid waste simply found its way downhill into rivers or straight into the sea. In small amounts, waterborne bacteria can reprocess organic effluent and convert it into harmless products, but when

NATURE'S POLLUTION MONITORS

Lichens can live on the most unpromising of surfaces – such as bare brick and stone – but they are highly sensitive to the state of the air. During the 19th century many urban lichens in Europe and North America started to disappear, because they could not tolerate the sulphur dioxide that was created by burning coal. Since then sulphur dioxide levels have dropped, and lichens have staged a recovery. Some now flourish even in busy city centres.

thousands or millions of people live side by side this system of natural purification begins to break down. London's Thames was a prime example of what often happened as a result. Around 1800 its water was so polluted that one member of Parliament wrote a letter to the Prime Minister complaining about its appearance and smell. He used the river water as ink.

Polluted city air is a problem with an equally long history, especially in places where heating is needed for winter warmth. In London and other large cities of northern Europe, the situation was bad enough when wood was the main source of fuel, but when wood supplies started to dwindle and

coal replaced them – from about 1600 onwards – atmospheric pollution reached new heights. In 1808 the English poet Robert Southey wrote that London's air was 'a compound of . . . chimney smoke, smuts and pulverised horse dung', and this blanket of pollution blackened everything that it touched. Wealthy city-dwellers often had their houses repainted every two years, and St Paul's Cathedral, whose dome towered over the city, was caked with soot even before it had been completed. Given this kind of pollution, it is hardly surprising that respiratory problems were rampant.

Although these cities were still small by modern standards, they had already started to have a noticeable effect on their surroundings. London's air poisoned lichens that lived on trees downwind, and its waterborne pollution poured downstream like a deadly plume. By 1850 the fish that once thrived in the Thames's lower reaches had completely vanished.

A CHANGING IMPACT

Since that time many things have changed. First and foremost, towns and cities have continued to become bigger and more numerous, and they now house a much greater proportion of the world's population than they did before. In 1850, when the world's head-count stood at about 1.2 billion, less than 15 per cent of the population lived in urban areas. Today, with the population more than four times that size, half the world's people live in towns or cities, but, because this is an average, in many countries the proportion is higher still. In North America and much of Europe, about four-fifths of the population live in built-up areas. In Australia – an island continent that paradoxically has vast areas of open land – the figure is nearly 90 per cent.

Statistics like these show that city life is fast becoming a hallmark of the human race.

But although cities have grown rapidly in the last century, their impact is not the same as it once was. In the industrialised world at least, some of the problems of the past have been successfully dealt with, although new ones have often taken their place.

The River Thames shows what can happen when urban pollution is tackled in a coordinated way. Its steely grey waters now flow through a city of over 6 million inhabitants and out to sea through an estuary flanked by housing and factories. However, beneath the river's unpromising looking surface, an extraordinary change has taken place. The river's oxygen content – a measure of its biological health – has climbed from almost zero in the 1950s to about half the possible maximum, and a river that was once biologically dead has come back to life. Dramatic confirmation of this turnaround came in 1974, when a salmon was discovered in the intake screen of a London power station. During medieval times salmon was a staple food of London's poor, but this solitary fish, perhaps making its way upstream in an attempt to breed, was the first to be seen in the Thames for 140 years.

The Thames clean-up has been echoed in other urban waterways, from the Rhine in Europe to the Hudson in North America. New York's Hudson River reached its biological crisis in the 1960s, and once again concerted action brought it back from the brink. In most cities in the developed world, effluent discharge from homes and industries is now much more carefully controlled than it once was, even though keeping rivers clean sometimes means disposing of waste somewhere else.

When the spotlight moves from water to urban air, the picture becomes more confused, and the gains and losses become more complex. Readily available electricity and tighter air quality regulations now mean that, in many northern cities, murky winter air, with clouds of soot blowing close to the ground, is largely a thing of the past. However, as many of today's city-dwellers are only too aware, urban air pollution has not gone away; instead, one form has often been replaced by another. Today's pollution comes not from domestic fires or the locomotives that once ferried people to work, nor from the horse dung that troubled Robert Southey, but from the exhausts of innumerable cars.

CHEMISTRY IN THE STREETS

Compared with smoking chimneys, car exhausts can look innocuous enough. Although they release a clutch of different gases, together with microscopic particles of partially burnt carbon, these are mostly colourless, and the carbon particles – unlike the soot particles of bygone days – are normally too small to be seen. But when the weather is calm and sunny, and when tens of thousands of cars are on the move, the energy in sunlight acts on exhaust gases and makes them react. The result is the aerial signature of many modern cities: a blanket of yellowish-brown smog.

continued on page 60

THE ULTIMATE CITY LIMIT

How big are the world's cities, and how much are they likely to grow? One way to find out is to measure their area, but cities vary enormously in density. Some consist of housing that is tightly packed, while others are mainly sprawling suburbs. A more meaningful figure is provided by the number of people they contain, but even here measuring cities is not always as straightforward as it sounds.

In 1984 a UN population conference took place in Mexico City, which was then estimated to contain about 17 million people. According to experts at that time, Mexico City's population was accelerating rapidly, and would reach 25.6 million by the year 2000. Other cities throughout the world were predicted to follow suit, with many having populations around the 20 million level by the year 2015. However, in the years since 1984 many of these cities have failed to follow their expected course. Mexico City had only 15.5 million ten years later, and several other cities have mysteriously 'shrunk'. There are two probable explanations for this disparity between prediction and reality. One is that precise population figures for large cities are difficult to obtain. Another is that predictions that simply extrapolate what is happening at present often come unstuck. At times in the past – particularly during the Industrial Revolution – some cities expanded at breakneck speed, but eventually their rate of expansion slowed. Today many of these cities are hardly growing at all.

Instead of expanding indefinitely, today's urban centres will probably follow the same pattern. When cities grow to very large sizes, the number of people moving out often begins to offset a city's population growth. When the two balance, the expanding city finally comes to a halt.

UPWARD GROWTH *The Bolivian capital La Paz was originally founded in a deep canyon offering shelter from mountain winds. Today its sprawling suburbs perch high above it.*

GREEN MACHINES

After decades of unrestrained growth, pollution from urban cars seems to have reached the point where people are prepared to tackle it in a determined way. There are two overall ways of doing so: one is to reduce the amount of air pollution that each car creates; the other is to reduce the number of cars.

If gasoline were a pure fuel, and cars burnt it completely, car exhausts would contain only carbon dioxide and water vapour. Unfortunately, oil-based fuels are complex mixtures of different substances. As well as hydrocarbons, which provide energy, they contain compounds of other substances such as sulphur and nitrogen. When these are burnt, they produce gases that dissolve in atmospheric moisture to produce powerful acids. One way to reduce the unwanted side effects is to use a catalytic converter. This functions as a kind of chemical work surface on which exhaust gases are reprocessed before they escape into the air. Another way is to use a completely different fuel. Ethyl alcohol can be produced from sugars formed by a wide range of plants, including sugar cane and sugar beet. Unlike fossil fuels, ethanol,

as it is called, is chemically pure, and because it is a single substance it burns very evenly, producing just carbon dioxide and water.

A more radical method of curbing pollution sidesteps combustion altogether. Until

GROWING FUEL *In the Netherlands 'green buses' are powered by ethanol, an alcohol produced from plants.*

now electric power has played a very modest part in road transport, but as concerns about pollution grow it looks set to become much more significant. However, unlike gasoline or alcohol, electricity cannot be stored; instead it has to be created as it is needed, usually by chemical means. One way to do this is to use a battery, but this adds a considerable weight to a car. Another way, which is already used in space, involves using fuel cells. These simple but ingenious devices combine hydrogen and oxygen, and produce an electric current; the only waste product is water.

The other method of pollution control – reducing car numbers – is simply a matter of planning. Many cities are now reinvesting in public transport systems, and some have

RAPID TURNAROUND *In Brazil, Curitiba's bus-stops help to speed up public transport. Passengers get on and off through raised doors, and buy tickets at the stop.*

THE TRAM RETURNS *Once seen as outdated, trams and light rail systems are again proving their worth in city centres.*

been designed to make travel without cars as easy as possible. Foremost is the Brazilian city of Curitiba, where fleets of specially de-signed buses, some capable of carrying 270 passengers, speed along express roads that are closed to other traffic. Curitiba's transport system has attracted attention from planners all over the world, because it has developed at a fairly low cost, is heavily used and includes many ingenious features. Among these are specially designed stops in which passengers pay before boarding. The result: instead of wasting time at stops, each bus spends most of its time on the move.

SOLAR CARS *Streamlined to reduce its air resistance, a solar-powered car travels by the energy of the Sun.*

This photochemical smog affects cities to different degrees, and the difference is not always due to the number of cars being used. Cities with a cloudy and breezy climate escape fairly lightly, but ones in sunnier parts of the world, where the air is often calm, are much more at risk, because here the gaseous blanket can linger for days at a time. Los Angeles and Mexico City are

TAINTED AIR *Photochemical smog hangs over the centre of Mexico City. Its characteristic colour is produced mainly by nitric oxide, a poisonous gas with a choking smell.*

famous for their smog, as are Bangkok, Jakarta and many other expanding cities in the developing world. But even relatively small cities, such as Granada in southern Spain, suffer from smog because their surrounding mountains cradle the air while the sunlight sets to work.

Like coal smoke before it, this smog not only makes life uncomfortable for humans – it also affects other forms of life. The toxic gases in smog, which include ozone, cause significant damage to plants, and in southern California they particularly harm those that evolved in the once-pristine desert air.

The way people travel affects more than city air. Over the centuries various different

methods of transport have helped to shape the cities we see today, and this in turn has fashioned the impact that they have on their surroundings. In all early settlements movement was on foot or horseback, so people had to crowd together to be near their work. By today's standards these cities were extremely compact, and even the largest of them could be crossed during the course of a brisk walk or ride. Instead of gradually dissolving into suburbs, they stopped with startling abruptness, giving way to open farmland beyond.

With the invention of passenger railways in the 19th century this pattern began to change. People could live much farther from

the city centre while still being able to travel to work. This era saw the birth of the commuter – a new word for travellers who exchanged or 'commuted' their daily tickets for season tickets – and it also saw the beginning of modern suburbs, where city and countryside met and merged.

The effect of this change was dramatic, as commuter belts developed like fingers stretching out from city centres. Observing it in southern England, the writer H.G. Wells saw clearly where it might lead. In 1902, just 17 years after the German engineer Karl Benz tested the first practical automobile, Wells wrote that a city that relies on horse-drawn transport could perhaps be stretched to a radius of about 8 miles (13 km), after which lines of communication would begin to break down. But the steam and internal combustion engines were changing things considerably, Wells asserted. Was it too much to expect, he asked, that 'the great city of the year 2000, or earlier, will have a radius very much larger than even that?'. For himself, he was sure of the answer: by the end of the millennium, commuters who could afford the tickets might be making return journeys of 200 miles (320 km) in a single day.

At the time it would have been an audacious statement, but today it is an accepted fact. Although not everybody would care to travel this far, this kind of journey is well within the bounds of possibility. The effect, just as Wells predicted, has been a 'centripetal pull' – a flinging outwards of urban development so that, instead of being firmly separated, cities and their surroundings are bound ever closer together. Cities are becoming more diffuse as people migrate away from city centres and live closer to the surroundings in which their city stands.

Modern city centres often live up to their reputation as concrete jungles, where

FORGOTTEN CITIES

A few of today's cities – such as Damascus in Syria – have been continuously inhabited for more than 4000 years. But during the course of human history many more cities have been abandoned, perhaps as a result of climate change, soil erosion or war. In the Middle East, where many early cities arose, it is relatively easy to spot crumbling remains in their treeless surroundings. In other parts of the world some major cities have disappeared beneath a cloak of vegetation, to be largely forgotten and then discovered many centuries later by outsiders. The city of Angkor Thom, in modern-day Cambodia, was the capital of the Khmer Empire. Founded in about AD 800, it was abandoned in 1434 and, although accounts of it were sent to Europe by missionaries in the 17th century, its jungle-clad ruins were only revealed to the outside world in the 19th century. In the Americas the Mayan city of Tikal lay almost untouched for 1000 years, despite the fact that its pyramids soar high above the forest

ANGKOR THOM *Angkor was at the heart of a state that covered lowland Cambodia. With the forest cut back, its magnificence is again revealed.*

that once smothered its site in modern Guatemala. In Peru the relatively recent city of Machu Picchu

had an even more remarkable history. Constructed by the Incas as a last stronghold against the Spanish invaders, it was completely unknown until its discovery by Hiram Bingham in 1911. Slung between two mountain peaks and covering about 5 sq miles (13 km²), it was perhaps the greatest lost city of all time.

all traces of the natural landscape have been covered up or stripped away. However, taken as a whole, many of today's cities are not like this. In a typical metropolitan area in a developed country, trees actually take up more space than buildings, and the patchwork of buildings and open land creates a habitat that is waiting to be used. Some forms of wildlife find it difficult to exploit, but for others the city and its fringes have much to offer.

STREET LIFE

All around the world a select band of plants and animals have successfully adapted to this urban world. Many of the plants grow from wind-borne seeds, and they manage to spring up between the cracks of paving stones, on waste spaces and vacant lots, and even on buildings themselves. Once they have taken root, some species – for example the creeping thistle (*Cirsium arvense*) – develop stems that spread horizontally

underground. Even when they are covered by asphalt, they can sometimes manage to punch their way to the surface. Plants like this often spearhead nature's transformation of derelict urban land, helping to turn bare ground into a mass of greenery during the summer months.

Walls may not look very promising places to grow, but if they are built of brick even they can provide plants with a toehold on busy city streets. Mortar contains minerals that plants normally get from soil, and water trickling from gutters supplies moisture. Given these raw materials, as well as air and light, some plants survive almost as well on walls as they would on the ground. A few can even become a problem, because their roots may grow big enough to create cracks between the bricks.

Some animals – such as the city pigeon – have become so successful in cities that they are now more common in these artificial surroundings than in their original homes.

FAST FOOD *In an American
backyard, a well-fed raccoon
sorts through the contents of a
dustbin in search of a meal.
Usually nocturnal, this raccoon
has ventured forth in daylight.*

SUBURBAN SANCTUARY *Seen
from the air, a Californian
housing estate is a patchwork
of trees and buildings –
a habitat for urban wildlife.*

But not all urban wildlife is as easy to see,
because many city animals are nocturnal.
This characteristic neatly dovetails with the
habits of the human population and gives
them a chance to live relatively undisturbed.

These night-time urban dwellers include
brush-tailed possums in Sydney, foxes in
London, raccoons in New York, and in some
American cities a bird called the common
nighthawk (*Chordeiles minor*). This small
insect-eater normally lays its eggs directly on
bare ground, using its dappled dark brown
plumage to hide from its daytime enemies.
When flat gravel roofs became widespread
during the last century, it was quick to adopt
them as a useful alternative, and roofs now
provide thousands of night-
hawks with a home.

Most of these urban animals
are full-time residents, although
the common nighthawks fly
south in the autumn when
the supply of airborne insects
vanishes. Often taking the nighthawks' place in the northern
cities is another set of travellers
of a very different kind. Instead
of heading out of town they
head towards it, and instead of
travelling just once, they set off
every afternoon.

Initially these airborne commuters travel in small groups,
but gradually the groups begin
to converge, until thousands
or even tens of thousands of them are on
the move together. They scud over roads
and rooftops oblivious to the traffic below,
and press on in a chattering mass towards
the bright lights in the distance. Their goal
is usually the city centre, and their motive
for reaching it is not food, but warmth.

This late afternoon invasion of starlings,
which takes place in many large cities in
Europe and North America, provides living
evidence for what many people instinctively

feel when travelling between cities and the countryside around them. Cities generate their own climates, and the conditions in a city centre are often subtly different from those farther out. For birds such as starlings the countryside usually offers the best prospects for finding food and raising young, but on winter nights city centres beckon with a warmth that can be difficult to resist.

ISLANDS OF HEAT

How different is the urban climate from the one around it? Research shows that it depends on the size of the built-up area, and also on the density of the buildings within it. Concrete, bricks and asphalt act as heat stores, soaking up warmth during the day and releasing it after dark. As well as these shifters of heat, cities also contain heat sources. These include the engines of cars, heating systems within buildings, and almost all forms of electrical equipment, from computers to vacuum cleaners. Some of these sources are very important: in Manhattan, during winter months, up to twice as much heat can come from combustion as from the Sun.

Added to these are biological sources of heat. With body temperatures of 37°C (98.6°F), we are usually much warmer than our surroundings, and each of us sheds heat in various ways. About 60 per cent of the heat is lost as radiant energy, which consists of heat waves that spread out all around us. About 35 per cent is lost either by convection, which warms the air around the body, or by making sweat evaporate. The remainder leaks away through conduction – as happens, for example, when bare fingers touch a cold doorhandle on a frosty day. Individually, the heat output of a resting adult is

much less than a domestic heater, but where thousands or millions of people are close together, its impact becomes more significant.

The relative importance of these different heat stores and sources is difficult to gauge, but their combined effect is simple to measure. On an annual basis, the centre of Berlin, London or New York is about 1°C (1.8°F) warmer than the city outskirts, but this relatively low average masks much bigger differences at certain times of the year and at particular times of day. In some North American cities the 'heat island' effect on winter nights can be so great that the city centre is more than 10°C (18°F) warmer than the land outside. Interestingly, European cities show less marked heat islands, probably because their city centres are not so densely built up.

These temperature differences have some intriguing effects. They help to explain why cities often feel hotter than visitors expect, and why plants in cities usually produce leaves and flowers earlier in the year than their rural counterparts. Urban heat also helps some plants to survive winters that would normally kill them off, and it has the same beneficial effect for a range of cold-sensitive animals, ranging from butterflies and ants to spiders and scorpions.

Temperature is only one factor that helps to make up a region's climate, and cities affect more than this. In a large city tall buildings stand in the way of moving air, and can sometimes cut the annual mean wind speed by about 30 per cent. Calm air is up to 20 per cent more frequent in city centres than in the open country outside, but the city climate is not always more tranquil as this suggests.

When buildings and roads start to release their stored heat, they produce warm air that rises far above the rooftops. This air usually contains moisture and tiny specks of dust, and as it rises it starts to cool and expand. The dust provides nuclei for condensing rain or hail, and if conditions are right, this suddenly spills out of the clouds, pouring onto the ground downwind. For most cities, this storm-generation is a fairly minor effect, but in a few – such as Houston and New Orleans – hailstorms are up to four times more likely within the city limits than outside them. This capacity to alter the weather shows that, as in many other areas, cities often have an impact that no one could foresee.

NATURE AT BAY *A Japanese shopping mall teems with people. Few kinds of wildlife are able to survive in these artificial surroundings.*

CHANGING HARVESTS

Ever since farming began, growing food has played an essential part in human life, and today more food is being produced than ever before. But as farming has changed and spread, so has its impact on the Earth's natural landscapes.

A farmer casts his eyes along a row of plants, and then walks down the row to see if the crop is ready to be picked. As always, the yield promises to be excellent. Nurtured by constant warmth and almost endless light, the plants have grown rapidly, and their deep green leaves show that they are in perfect health. But it is still early in the season, and a check shows that the crop is not quite ripe. It will be another few days before the pickers can set to work.

Scenes like this have been part of farming life for centuries, but what makes this one unusual is its location, and the way the crop has been grown. The plants are tomatoes, and they have been raised in the windswept and treeless landscape of Husavik on the north coast of Iceland. Beyond the sea's often stormy horizon, the Arctic Circle lies less than 50 miles (80 km) away.

With summer temperatures averaging a bracing 11°C (52°F), Iceland supports grazing livestock but is too cold for almost all crops except grass. Yet by harnessing volcanic heat and using

PROVIDENT PLAINS *Two centuries ago this area of North Dakota was a sea of short-grass prairie.*

PAMPERED PLANTS
In Iceland long summer days and abundant geothermal heat allow a wide range of exotic crops to be grown.

glasshouses to contain it, Icelanders successfully grow many foodplants – including aubergines, capsicums and even bananas – that came originally from much warmer parts of the world.

The existence of such tender plants in this cold part of the world is a testament to human ingenuity and perseverance. However, Iceland is not the only place where agriculture has managed to push back the limits that nature once imposed. Throughout the world, from the subarctic to the tropics, modern farming has transformed the art of growing food, as well as many of the places used to grow it.

BREAKING NEW GROUND

In the 1830s an American blacksmith created a piece of machinery that was to have a dramatic effect on the world's food production, and on some of its landscapes. The man was John Deere, and his invention was a new kind of plough.

Deere lived in Illinois and spent much of his time repairing ploughs, which made him keenly aware of their strengths and weaknesses. During his time ploughs were made of cast-iron parts bolted to a wooden frame. They had a small vertical blade called a coulter, which sliced through the ground, and a large plate or share that followed it

and opened up the cut. A separate flap, called a mouldboard, ran alongside the share, where it lifted the soil and turned it over to create a furrow.

By the beginning of the 19th century the mouldboard plough had existed for more than a thousand years, although it had seen a host of minor improvements. It worked well enough, even if its progress – pulled by a horse and creating just one furrow at a time – was painfully slow. However, when American farmers moved west and tried to till the soil of the Great Plains, their ploughs did not even move slowly. They more often came to a halt.

The problem lay not so much in the plough itself as in the ground beneath the farmers' feet. In this vast region of prairie grassland the soil was bound up by a dense network of fine, fibrous roots which penetrated far beneath the surface in their quest for moisture. They made the soil so firm that it could even be dug up in blocks and used for building. Confronted by this kind

of ground, traditional ploughs stuck fast and often broke apart.

Deere's answer to the problem was to turn the plough into a much tougher piece of equipment. In his version the share and mouldboard were of steel rather than cast iron, and they formed a single wedge-shaped unit rather than being two separate parts. Modified in this way, Deere's plough could gouge its way through the black prairie soil, allowing the ground to be planted for the first time. It also made ploughing more efficient in fields that had already been tilled, because less work was needed to drag it through the ground.

The impact of Deere's innovation took time to assert itself, but it would be phenomenal in its extent. Cultivation spread outwards from the eastern edge of the Great Plains, and crept inexorably across the gently rising ground towards the Rocky Mountains in the far west. In the early 19th century the total area under crops was almost too small to quantify, and was of little importance compared with the farmland farther east. But by 1900 it had already leapt to 75 million acres (30 million ha), and when tractors were introduced after the First World War it continued to surge, reaching over 150 million acres (60 million ha) by 1980. The open prairies, with their distant horizons of almost limitless grass, became a thing of the past.

RITE OF SPRING *A 19th-century plough cuts a furrow through an English field. Before mechanisation, growing food required large amounts of labour.*

SOIL ON THE MOVE *A 1930s dust storm looms over fields in the Texas panhandle. Storm fronts like these could easily overtake a moving car.*

The great ploughing-up that took place in the United States and southern Canada also swept across grasslands in other parts of the world. These included the pampas of Argentina, parts of Australia and, more recently, parts of the former Soviet Union. It took little more than a century and was one of the swiftest agricultural changes the world has ever seen.

THE GREAT EXPERIMENT

When any region is opened up to agriculture for the first time, farmers become players in a great outdoor experiment whose outcome depends on many factors. Some take effect straight away, but others – as the Great Plains farmers discovered – can remain hidden for many years before they reveal their hand. As they cut the first furrows through the ground, the pioneer farmers had good reason to be optimistic. The soil was very fertile, and wheat in particular seemed to thrive in the newly broken ground. Harvests boomed, and the prospects for the future looked good.

In the 1890s their good fortune was suddenly checked. A short but severe drought stunted the crops and provided the first hint of the kind of trouble that might lie ahead. Wheat farming soon recovered from this temporary setback, and the ploughs continued their westward progress. But a generation later, during the 1930s, the climate dealt a far more deadly blow. This time its effects could be seen not only on the ground but also – and far more frighteningly – in the air above.

Dust storms are a common enough occurrence in the world's deserts, but in farming country they have frightening implications. For day after day between 1934 and 1938, winds powered by advancing weather fronts swept up the prairie soil into airborne banks as much as 2 miles (3 km) high, and simply blew it away.

Today, over 60 years later, the worst storms of the 'Dust Bowl' years almost defy the imagination. Although the storms principally affected the Midwest, their influence was felt much farther afield. According to one report, a four-day storm in May 1934 'transported some 300 million tons of dirt 1500 miles [2400 km], darkened New York, Baltimore and Washington for five hours, and dropped dust not only on the President's desk in the White House, but also on the decks of ships some 300 miles [480 km] out into the Atlantic'. In the same year 300 dust storms swept across North Dakota in eight months, while in 1936 the region that straddles the border between western Oklahoma and northern Texas endured 22 days of dust storms in the month of March alone. With the topsoil stripped away and the air burdened with dust, farming was quite impossible.

What had gone wrong? Why had the once-productive land turned against those who had settled it? The answers lay in the unintended effects of the plough.

In the original prairie landscape – particularly in the drier west – the network of grass roots acted like a living blanket, and held the soil together even in the severest droughts. But once the roots were broken up by the plough that skin was torn apart. The root fragments began to break down,

WOUNDED EARTH *Gulley erosion was widespread during the Dust Bowl Era. Once the soil was stripped of its cover, sudden rainstorms could cut deep gullies into its surface.*

and after several decades of cropping the soil consisted mainly of dust-like particles of rock. The stage was then set for a crisis. When weather patterns changed and prolonged drought set in during the 1930s, the wind scooped up the dust and forced many farmers to pack up their belongings and abandon the land that once supported them.

THE BIRTH AND DEATH OF SOIL

As each year goes by, fewer people are left who recall those nightmarish scenes at first hand. But as farmers continue to work the Earth's surface, the problem of soil erosion is still very much with us.

Soil is a remarkable substance, created partly from weathered rock and partly from the remains of living things. Although it is a renewable resource, and is constantly built up and worn away, nature renews it extremely slowly indeed.

To find out just how slowly, geologists look to places where completely bare rock has been exposed from a point that can be accurately dated. One such place is Alaska's Glacier Bay, where periodic changes in climate have made ice fronts advance and then retreat in their progress towards the sea. Here studies have shown that when bare rock is revealed by the ice, about 6 in (15 cm) of soil forms in the first century. After this the increase slows and finally stops altogether, as formation and erosion begin to balance out.

This rate of soil formation may sound fairly sluggish, but in fact the reverse is true. The Glacier Bay area is soaked by heavy rain blowing in from the Pacific Ocean, so its bedrock weathers at an unusually rapid rate. In most of the temperate world rainfall is much lighter, and so soil forms at a more leisurely pace – typically less than 1 in (2.5 cm) a century. In regions that are drier still, soil formation is even slower. Here the depth of soil built up in an entire human lifetime is sometimes not

SCOURED BY ICE *When glaciers grow they bulldoze the ground free of soil. When they retreat, the process of soil formation begins all over again.*

much thicker than a sheet of paper.

With these slow rates of renewal, human interference with the soil can have profound effects. The awe-inspiring dust storms of the 1930s may have passed into history, but across much of the inhabited world topsoil is still being lost much faster than it is being formed. In the United States alone, 75 billion tons of it are washed or blown away each year, which is equivalent to the weight of an average man for every 25 sq yd (21 m^2) of cultivated land.

Left unchecked, this kind of change in the Earth's surface has the potential to trigger a catastrophe because, without the soil, cultivation of any kind is impossible. But while soil erosion can never entirely be stopped, it can be greatly reduced. One of the oldest techniques for combating it appeared more than 7000 years ago, while one of the newest has become widespread only in the last two decades.

KEEPING THE LAND IN PLACE

Although it far predates any written records, the origin of terracing is not hard to visualise. Imagine, for example, a farmer surveying the aftermath of a short but violent downpour as he trudges over his gently sloping land. As he checks the ground, he encounters an unwelcome sight – gulleys reaching uphill like grasping fingers, and streaks of bare rock showing where the thin topsoil has been scoured and then washed away. But in one spot, where a line of boulders stood in the path of the escaping water, something interesting has happened. Here the rocks have held back the flow, and a layer of damp soil now glistens in the sunshine. The farmer looks at it, and perhaps tests it with a stick or a probing foot. The soil is several inches deep and contains no stones. Seeing an opportunity in this accident, the farmer decides to imitate nature. He rolls more rocks into position, until eventually a low wall runs at right angles to the direction of the slope.

From such simple beginnings, the art of terracing has grown into one of mankind's most important transformations of the natural landscape. On low slopes terrace walls need not be much higher than steps, and making them requires little skill. But on steep slopes they have to be both taller and stronger, particularly if they are to hold back the water needed to grow rice. On mountainsides in South-east Asia, where terracing is at its most advanced and intensive, individual terraced fields may be only a few feet wide but their supporting walls can be over 15 ft (4.6 m) high. Without these walls, nearly all traces of topsoil would soon be washed way.

Maintaining this kind of landscape requires an immense amount of work, and is feasible only when land is in short supply and labour abundant. But the principles behind terracing still hold good even in very different landscapes such as the American

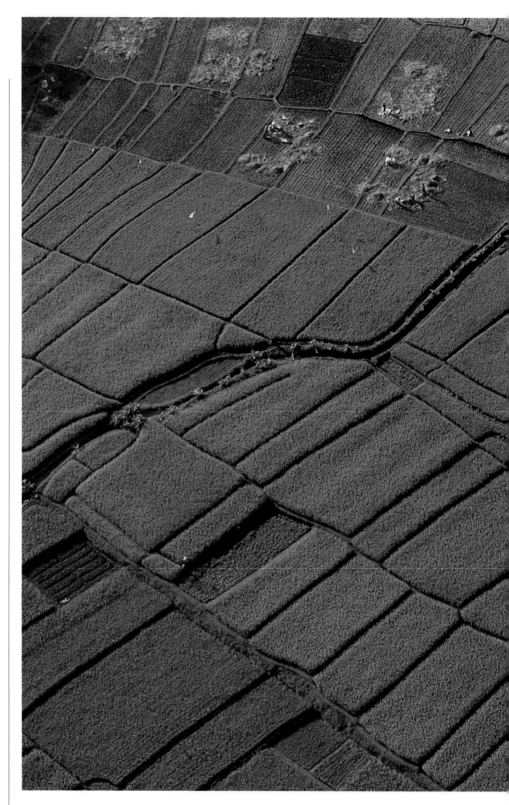

Great Plains. Here farmers on gently sloping land often plough along contours, rather than up and down slopes, creating miniature terraces that help to hold the soil in place.

Across much of the American wheat belt the land is so flat that there are no contours to plough along, yet erosion is still a problem. To combat it an increasing proportion of farmers have tried a new and radical technique: they have given up ploughing altogether. To the pioneer farmers of the Great Plains, growing crops without ploughing would have sounded like nonsense, but, once the ground has been opened up, no-tillage agriculture may be the key to farming's long-term future.

GREEN STAIRCASE *A flight of terraced rice fields drops down a steep hillside in Bali (left). In a vertical strip of fields the harvest has already begun. Contour ploughing (above) creates a landscape of abstract shapes in the Palouse country of Washington State.*

by as much as four-fifths – an important saving of a precious resource.

WATER ON THE LAND

As ancient as the plough is the device usually known by its Arabic name: the *shaduf.* It looks like a small crane, with a wooden beam that pivots on an upright post. A counterweight hangs from one end of the beam, while a bucket – originally made of leather, but nowadays more often metal or plastic – hangs from the other.

In the no-tillage system, farms still have their ploughs, but they are treated with the caution they deserve and brought out as rarely as possible. Seeds are injected – or drilled directly – into the ground, often underneath the previous crop's unploughed stubble. Instead of turning over the soil and leaving it exposed to wind and rain, this system preserves the ground's organic skin, making it much more difficult for the soil to become dislodged.

In just two decades, no-tillage agriculture has proved to be a powerful weapon in the fight to conserve precious soil. In western Canada studies of similar plots of land have shown that it can reduce soil erosion

When in use, the *shaduf*'s movements follow an easy and graceful rhythm. The farmer pulls down on the empty bucket, raising the weight on the other end of the beam. The bucket drops into a river or ditch and fills with water, at which point the farmer gives a light tug. The weight then falls and the bucket rises, the beam swivels on its post, and

DRAWING WATER *In an oasis
on the northern fringes of the
Sahara a double-beamed*
shaduf *helps herdsmen to raise
water for their livestock.*

once parched landscape grows a vast range
of fruit and vegetables, including moisture-
loving crops that would die within hours if
the water supply was cut.

THE HAZARDS OF IRRIGATION

While the technology of irrigation may have
changed, one of its unwanted side effects
has not. In irrigated drylands across the
globe, farmers still face an implacable problem that dogged their predecessors long
ago: the man-made threat from salt.

On its journey from the Himalayas to the
Arabian Sea, the River Indus passes through
a broad plain over 500 miles (800 km) long.
The rainfall here is low, and for months at a
time the plain bakes in temperatures over
35°C (95°F). Without human influence the
boundary between the river and plain would
be simple and stark, with the flowing water
bordered by two ribbons of green, themselves flanked by desert. But here, as in the
Middle East, irrigation has transformed the
landscape. During the last 100 years croplands have expanded rapidly, creating vivid
oases in the scorched landscape.

It sounds like an agricultural success
story, and in part – at least – it is. But success has been bought at a price. While the
area under irrigation is still growing, large
parts of it have already been abandoned.
Instead of growing crops these fields lie
idle, covered by a patchy crystalline crust
that glints like frost in the sun.

Scenes like this are not new, nor are they
unique to the Indus valley. Temple records
from Mesopotamia – home of the Sumerians,
and one of the first areas to be irrigated –
provide remarkable evidence of the insidious
effects of salt-laden soil. The records show
that from about 2500 BC harvests began to
decline and land fell out of production.
Originally farmers grew a roughly equal mix
of wheat and barley, but as yields fell they
concentrated on barley, which proved better at coping with the salty ground. However,

the liquid load is emptied into a channel
that runs away into a field. Armed with the
shaduf and other devices for raising water,
farmers in the Nile Valley and Mesopotamia
transformed their desert surroundings into
a lush patchwork of productive fields. The
irrigated ground was so fertile that it provided food in abundance, and this allowed
some of the world's first cities to flourish.
Today, several thousand years later, nearly
15 per cent of the world's croplands are arti-
ficially watered, and they produce a third of
the world's food. Without them we would
have great difficulty growing enough to eat.

In some places irrigation still depends
on simple pieces of machinery like the
shaduf, but in others water is distributed by
a complex network of pumps and canals
that carry it for hundreds of miles. Irrigation
can simply top up local rainfall, making
it easier to produce a reliable harvest, or it
can bring water to places where it is otherwise so scarce that no crops will grow. In
the Imperial Valley of southern California,
for example, rainfall is often less than 5 in
(12.5 cm) a year, which is only enough to
support the most resilient desert plants.
However, with the help of irrigation, this

even this measure did not prove to be a long-term solution, and in the space of seven centuries the amount of food harvested on each acre (0.4 ha) of ground dropped by two-thirds. Weakened by this inexorable decline, and under attack from outsiders, the Sumerian civilisation collapsed.

The threat from salinisation today is not quite as grave as this, but in a world where land is a valuable resource it is still a major problem. Salt has claimed farmland in places as far apart as California, Spain and Central Asia, and in Australia – the worst affected continent – it has eaten into more than 10 million acres (4 million ha) since European immigration started. In south-eastern Australia alone it currently costs about $75 million a year in lost production.

The farmers of Mesopotamia probably had little idea why salt was devastating their land, although they might well have guessed that it had something to do with

TRAVELLING SOIL

In order to remain stable, ships sometimes have to take on ballast to increase their weight. In northern Europe, Norwegian sailing ships returning home sometimes took on a very unusual form of ballast – soil. The soil was collected in the Netherlands and carried northwards to islands off the Norwegian coast. The islands' original soil had been stripped away by glaciers, but the Dutch soil enabled them to be farmed for the first time.

water. Today the forces behind salinisation are well known, and an ordinary indoor plant can help to explain one of them. A typical houseplant lives in a warm, dry atmosphere – much like that of many deserts – and it relies on watering, or irrigation, to survive. When the plant is watered, the

water is held by the soil, and as time goes by some of it is absorbed by the plant's roots. The rest slowly evaporates from the soil's surface, and it is this water that is at the heart of the problem.

Water is a superb solvent, dissolving a range of mineral salts that occur naturally in soil, but when it turns into a vapour it leaves all these salts behind. The result – in a flower-pot – is a harmless scattering of whitish spots. Scaled up countless times and transferred to hot dry farmland, it becomes the crystalline crust that is lethal to most plants.

RECLAIMING SALT-LADEN GROUND

During the last 300 years about a quarter of the world's irrigated land has been abandoned. At one time nearly all of this land would have been given up as lost, and nothing more done with it. But armed with a knowledge of the way salinisation occurs, today's farmers are in a much better position

WHEN ANIMALS CHANGE THE EARTH

Some of the most important changes created by farming are brought about not by growing crops but by raising animals. Light grazing often checks the growth of woody plants, and can encourage useful species, such as grasses, that would normally be shaded out. But if too many animals are raised on an area of land they can change it beyond recognition.

The aftermath of this process can be seen in places such as Greece and the northern Middle East. Here the land was once covered by extensive forests, which were gradually cut down by early farmers. The livestock these farmers owned – mainly sheep and goats – nibbled away at bushes and saplings, and prevented the vegetation from recovering. On the steepest slopes the plant cover became so sparse that the soil was washed away by winter rain, leaving the bare rock that

remains today. The landscape may look natural, but it is man-made.

Overgrazing can also cause deep gulleys that eat their way into level ground. In some parts of the tropics these gulleys can be over 40 ft (12 m) deep. As the rain erodes their sides, they can spread across pasture at the rate of many yards a year, swallowing trees and houses along with the soil.

With the invention of barbed wire, first produced commercially in 1874, overgrazing became a problem in different parts of the world. In America's rangelands, for example, it enabled farmers to build up cattle numbers, safe in the knowledge that

STRIPPED CLEAN *Sheep and rabbits have nibbled away the vegetation on a hillside in New South Wales after the natural tree cover was cleared. The soil has disappeared as a result.*

their animals could not wander off when food became scarce. Today, some parts of the American southwest – where water is often short – carry more animals than the land can

comfortably sustain. This overstocking already contributes to soil erosion; if continued for long enough, it may mean that the land is no longer able to support livestock at all.

to counteract it and to deal with its long-term effects.

To reduce salt build-up, water use has to be kept to a minimum. Open channels can often be replaced with a network of pipes that deliver a gradual trickle of water to individual plants, and thirsty crops can be exchanged for ones that are better at withstanding dry conditions. Even simple measures – such as watering plants by night instead of by day – can cut down the amount of water that is on the move.

To cleanse salt-laden soil, meanwhile, can involve going to the opposite extreme. The ground is flushed out with a large amount of water, which runs away into specially built drains. As the water runs away, much of the salt goes with it. However, even if the water were available, this kind of treatment is a luxury that not all farmers can afford. In Australia and North America several experimental techniques are being tested for reclaiming severely saline ground by planting shrubs and trees that are unusually good at tolerating salt. These soak up the water near the soil's surface, lowering the water table and reducing evaporation. If all goes according to plan, these plants will eventually be replaced by a mixture of trees, crops and grass, making barren land fertile again.

Despite the setbacks from salt and soil erosion, the world's croplands have doubled

LETHAL SALT *Dead trees and brackish water pinpoint salt-damaged ground near Swan Hill in south-east Australia.*

in little more than a hundred years. About 11 per cent of the world's land surface is now devoted to growing crops, while a larger proportion – perhaps as much as 30 per cent – is used for grazing livestock. But as agriculturalists survey the world today, opinions differ about how much room for expansion is still left. Some believe that the world's croplands could double again – pointing to deserts as the next farming frontier. Others think that, although we are

not yet running out of space, we may be reaching the limit of land that could prove economical to farm. If this is true, the future will lie not so much in expanding the area being farmed as in making existing farmland more productive.

BIGGER AND BETTER

The drive for increased yields is as old as agriculture itself, but until relatively recently it had little impact on the world beyond the farm gate. This was because farming was essentially a self-contained business, needing little or nothing from outside. Farmers kept the land fertile mainly by using manure from their own animals, by rotating their crops,

and by ploughing in the unused remains of the harvest. Even the seed they used was often collected on the farm itself. But in some countries, at least, this situation was not to last. The change began in the early 19th century, and it had its origin in one of the most unlikely places on Earth, where any form of farming is completely impossible.

Along the bleak desert coast of southern Peru one of the world's most bizarre agricultural commodities became the focus of major international trade. Known originally by its Peruvian name *huanu* – which became *guano* in Spanish – it consists of sea-bird droppings laid down in deposits up to 180 ft (55 m) deep. These deposits formed on offshore islands where birds have nested for millennia, and they built up to such immense depths because in the extremely dry climate there is no rain to wash them away.

Guano proved to be one of the world's most concentrated natural fertilisers. It is rich in nitrogen, phosphorus and potassium – three chemical elements that are essential for plant growth – and its discovery in such huge amounts triggered the organic equivalent of a gold rush. Between 1840 and the late 1870s the guano islands were the scene of intense activity, and over 10 million tons of guano were excavated, loaded onto sailing ships and carried away to Europe or the United States. Unlike cattle or horse manure this wonder substance had a truly remarkable effect on crop yields, making it worth transporting halfway around the Earth.

By a striking coincidence – engendered partly by the intensely dry climate – guano was not the only natural fertiliser that this part of the world produced. In 1809 large deposits of nitrate rock were discovered in the region that straddles Peru and northern Chile, and

FORTUNES FROM AFAR An abandoned guano loading platform survives on the desert coast of Peru. The guano boom made fortunes for dealers in this unusual fertiliser.

this also became a commodity in which vast fortunes were made. Unlike guano, nitrate had a double attraction: when crushed and added to fields it provided plants with nitrogen for growth, but it also could be used for making explosives.

For those who could afford them, these exotic fertilisers had a profound effect on farm productivity. However, a much more significant breakthrough came in 1908, when the German chemist Fritz Haber discovered a way of chemically combining or 'fixing' nitrogen from the air to make ammonia. From ammonia it was a fairly easy step to produce the nitrates that had previously been yielded by guano or rock. This was a momentous development, because nitrogen makes up nearly three-quarters of the atmosphere and is therefore practically inexhaustible. For the first time, large-scale production of synthetic fertilisers became a commercial possibility.

The results, as with John Deere's work on the plough, began slowly, but gradually gathered pace. In 1930 the annual global output of nitrogen fertiliser was about 1 million tons, and by 1960 it had reached more than 10 million. By 1990 it was more than 80 million tons, and is still climbing rapidly.

Meanwhile, along the coast of South America, the guano industry collapsed. However, even nowadays it is not entirely dead. As long as the sea birds continue to nest, guano still accumulates at the rate of

about 3 in (8 cm) every year. With careful management, this curious resource could be harvested indefinitely.

THE NEW PLANTS

If a group of 19th-century farmers could somehow be brought forward to the present day and dropped into the middle of a ripening field of wheat, they would have many reasons to be astonished. One would be the field's size, dwarfing anything that could be tackled with the power of the horse. Another would be the lack of any visible weeds or pests. But the greatest source of wonder would probably be the wheat itself. The packed ranks of almost identical plants, each laden with heavy ears of grain, would seem like a vision of almost miraculous plenty.

This impression of abundance is borne out by some remarkable figures. Throughout the world, wheat production has risen threefold in the last 50 years. In the United States maize yields have risen by up to eight times since the 1930s, and in neighbouring Mexico they have quadrupled since the 1950s.

UP AND COMING *The winged bean is one of several crops that may play a bigger part in world food production. Its pods and roots can be eaten, and like all members of the pea family it adds nitrogen to the soil.*

ENERGY FROM FARMING

Modern farming is highly efficient and uses very little labour, but its energy yields can be surprisingly low. When fuel and fertilisers are taken into account, North American wheat yields about 1.7 times the energy that it takes to cultivate. Tomatoes yield about half the energy that is spent on them, while spinach yields only a quarter. By contrast, peasant farmers in Mexico, working without tractors or fertilisers, can produce 10 times the amount of energy that they put into cultivating their staple foods.

Artificial fertilisers have certainly played a part in this explosion of agricultural production, as have the herbicides that eradicate competing plants and pesticides that destroy a crop's animal enemies. But these agricultural chemicals, which were all unknown 100 years ago, are only a part of the equation. The rest is made up by the 20th century's greatest advance in improved productivity – the transformation of crops themselves.

To understand what has happened to crops like wheat, imagine a farmer stepping forwards in time being replaced by one stepping back. Apart from the weeds, the insect pests and the eerie lack of motor noise, today's farmer would be struck by the height of the crop (old wheats were taller than modern types) and also by its diversity. A century ago almost every characteristic of the plants, from the number of their stems to the size and shape of their ears, was subtly but almost infinitely varied.

The explanation for all this diversity lay in the way the crop was produced. At harvest time a proportion of the seed would be set aside to sow the following year's crop, or exchanged with a neighbour. The result, over a long period of time, was that crops evolved into local varieties called landraces. Each landrace had recognisable overall features – some good, others not so beneficial – but it

was impossible to predict what kind of plant would grow from any single seed.

In the year 1900 a unique coincidence heralded the beginning of a far more precise way of breeding crop plants. Working independently, three European scientists – Hugo de Vries, Erich Tschermak von Seysenegg and Karl Correns – rediscovered research carried out by the Austrian monk Gregor Mendel over 40 years before. In the tranquil surroundings of a monastery garden Mendel had conducted breeding experiments on more than 20 000 pea plants in order to discover exactly how characteristics are handed on from one generation to another. Using techniques that have since become standard procedures, he painstakingly transferred pollen from flower to flower, and kept meticulous records of the characteristics that resulted in the next generation when particular parent plants were allowed to breed.

Mendel's findings were startling. In his time, it was widely assumed that differing characteristics – such as the colour of flowers – simply blended together when they were passed on, rather like differently coloured paints being mixed. But Mendel's pea plants showed that this was not so. Instead of blending, they usually remain quite separate, so that two pea plants, for example, one with red flowers and one with white, produced offspring whose flowers were also either red or white, but not pink. Mendel also discovered that these patterns of inheritance worked according to strict mathematical laws. The laws enabled him to predict exactly what proportion of seedlings would inherit a particular characteristic of their parents – something that had never been done before.

It was a brilliant achievement, but when Mendel published his work the scientific world paid little attention. As the abbot of a provincial monastery he was removed from leading academic circles, and his unassuming

FOSSIL FUEL *A combine harvester traces out a spider's web in a contour-ploughed field. Oil powers machinery, and provides the energy needed to make fertilisers.*

nature held him back when others would have pressed for recognition. However, by 1900 other minds had started to turn to the questions that Mendel had addressed. Although Mendel himself was now dead, his results – compiled with so much patience over many years – provided de Vries, von Seysenegg and Correns with valuable evidence to back up ideas of their own.

Once its true value was appreciated, Mendel's work had an almost incalculable effect. It spawned an entirely new branch of science – genetics – and has revolutionised our control over the plants we eat.

THE GREEN REVOLUTION

Much of the Earth's surface has a long history of cultivation, and our knowledge of how to use it has built up over many generations. But the potential that lies within plants is a much more recent discovery, and only during the 20th century has it been exploited in a systematic way. Two of the most remarkable examples of this can be seen in wheat and rice. Shortly after the Second World War the Wheat and Maize Improvement Centre was founded in Mexico, and in 1962 a similar foundation, concentrating on rice, was created in the Philippines. Together their scientists examined vast numbers of cultivated plants in farms, and combed remote areas in a search for wild plants closely related to these two crops. Though the work was complex, the goal was simple: to hunt down, collect and use any plants whose characteristics might help to improve crop yields.

For this enterprise to be carried out, some major practical problems had to be overcome. One concerned time. The life cycles of wheat or rice, although brief in human terms, still take several months to play through, and there is little a scientist can do to hurry them along. Before the result of any breeding programme could be assessed, many generations of plants often had to be raised, and, once a promising variety had been identified, another six or seven generations were sometimes needed to ensure that it produced consistent offspring. Saddled with this sort of schedule, the research threatened to take many decades.

Fortunately, human ingenuity provided some answers to this obstacle. In some parts of the tropics, good growing conditions allow two or even three crops to be squeezed into a single year, and using this principle the plant breeders concentrated on packing the generations back-to-back. With rice – originally a tropical plant – it was relatively simple, but in the case of wheat it meant growing the plants in places with very different climates – for example, warm lowland plains and cool mountain slopes – and then ferrying the seed from one to the other.

The results were outstanding. For both crops, scientists succeeded in creating new

SMALL-SCALE FARMING *Young hybrid rice plants soak up the morning sunshine in Sumatra. Farming on this scale still relies heavily on human labour.*

BREEDING RICE *A researcher uses a vacuum pipe to remove male stamens from a rice plant. Pollen from another plant will be used to create hybrid seeds.*

dwarf varieties that concentrate their resources into forming seeds rather than unwanted stems, and which do not fall over when burdened by their own grain. When raised in the right conditions these new varieties yielded unprecedented harvests, and in the 1960s and 70s a 'Green Revolution' spread throughout many parts of the developing world.

FUTURE HARVESTS

Whether it involves ploughing up new land or breeding new plants, the history of farming shows that new developments rarely have exactly the effects that are expected, and the Green Revolution was no exception. Yields have increased dramatically, but with this have come some far-reaching changes in the way land is worked.

When the first wave of new crop varieties was released, farmers in developing countries faced some difficult decisions. The new high-yielding plants needed generous amounts of artificial fertilisers, herbicides and pesticides to achieve their full potential. With this new and often alien form of husbandry they produced extraordinary yields, but without it they sometimes fared only marginally better than the varieties already being grown.

In rice-growing areas, in particular, farmers who were able to afford the extra

HOME-GROWN *Small farms, like these ones in Sri Lanka, still support a large part of the world's population. Increasing their yields will play a vital part in feeding the world.*

chemicals increased their harvests, and over the years became able to buy out their neighbours. Instead of helping everybody by providing more food, 'miracle rice' helped some but left others worse off than before. Today, the lessons of this paradox have been learnt and, instead of creating plants that need the care and attention that only high-intensity agriculture can provide, breeders now produce a range of varieties suited to many different kinds of farming.

The Green Revolution is still far from over. As recently as 1994 plant breeders succeeded in creating a new form of 'super rice', which produces seed heads on every shoot rather than on about two-thirds of them. If grown in the right conditions, this new variety has the potential to produce over 5 tons of grain per acre (12.5 per ha), which is nearly a third more than the best harvests at present. This new form of rice was produced by the techniques of plant breeding pioneered by Gregor Mendel, which rely on genes that are already present within an individual species. But with the advent of genetic engineering, which can transfer genes from one species into another, the stage is already set for an even greater wave of change.

THE LAND RESHAPED

Since the dawn of recorded history people have refashioned the Earth's surface. Grandiose monuments attest to the power of past civilisations, while even larger structures – roads, mines, canals, dams – reveal our greater capabilities today.

On the shores of Lake Van, lying amidst the harsh landscape of eastern Turkey, a mountain peak marks the final resting place of Antiochus I, a king during the 1st century BC. Two ranks of carved heads face the eastern and the western horizons, and a giant stone altar lies open to the sky. But the most extraordinary feature of this place is not the altar or the statues, or even the remarkably long inscriptions that were chiselled into them long ago. It is the mountain itself. Mount Nemrut has an artificial top, which is nearly 200 ft (60 m) high and covers an area of over 7 acres (2.8 ha). It was constructed from stones that were carried up the mountainside, and somewhere inside lies the tomb of Antiochus. Fired by their monarch's self-importance, the people of a minor kingdom wedged between Europe and Asia devoted millions of hours of labour to constructing one of the biggest ceremonial mounds outside the Americas.

This immense monument, built long before the days of powered machinery, provides striking evidence of the human ability to refashion the Earth's surface using muscle power alone. Further examples can be seen in other continents, some dating back before the beginnings of recorded history. England's Silbury Hill, a mysterious cone of chalk piled up in about 2700 BC, is 130 ft (39 m) high and weighs more than 0.5 million tons. Until the Industrial Revolution it was the biggest man-made structure anywhere in Europe. Monk's Mound, a flat-topped earth pyramid near St Louis, in the USA, is almost as tall as Silbury Hill, but it rises from an even bigger base. It was completed in the 12th century by the people of the so-called Mississippian culture, and once looked out over a densely inhabited landscape that was dotted with many smaller earthworks, some topped with houses, others used for burials.

Massive though it is, even Monk's Mound is dwarfed by a structure farther south. Outside the Mexican city of Cholula the brick pyramid of Quetzalcoatl, now partly covered by a blanket of earth and vegetation, sprawls over a base big enough to engulf a modern housing estate. This pyramid is much lower than the largest pyramids in Egypt, but its volume – over 4 million cu yd (3 million m³) – makes it the greatest structure of its type in the world. It was the outcome of many episodes of building and probably took at least 1500 years to complete.

From today's perspective, the motives that drove people to create these enormous structures are often difficult to interpret. But many other transformations of the Earth's surface, from ancient times to the present day, have much more practical aims. Natural landforms are not always ideal for human purposes, and nor are the conditions that they create, but instead of simply taking things as they are, like most other living things, we often set about changing them. With the help of excavators, bulldozers and dynamite, we now shift about 200 billion tons of the Earth's surface every year – enough earth, rock and rubble

THE WAY AHEAD *The Romans' Via Appia was one of the world's vital highways. Its rutted stone surface shows the effects of centuries of wear.*

to cover the entire state of New York in a layer more than 2 ft (61 cm) deep.

Of all the structures built by human beings, one kind above all has come to epitomise our increasing dominance of the planet. Whether they are vital arteries of commerce or the instruments of environmental change, roads play an inescapable part in the daily lives of the vast majority of the human race.

LINES ACROSS THE LANDSCAPE

The world's road network is beyond any calculable length. The United States alone has nearly 4 million miles (6.4 million km) of officially classified roads, and also contains part of the world's longest single highway, which runs almost without interruption from Alaska to Argentina and Chile. The world's highest road reaches an altitude of over 20 000 ft (6090 m) in Tibet, while the lowest drops to 1290 ft (393 m) below sea level as it

skirts around the arid shores of the Dead Sea. With the exception of extensive swamps, such as the one that blocks the Pan-American Highway between Panama and Colombia, hardly any land habitat has proved beyond the reach of this ever-growing network, and even tropical rain forests are being opened up by its radiating threads.

For most people today it is hard to imagine how life must have been before surfaced highways existed. But while tracks have a very long history, roads that could take vehicles were a much later invention. The ancient Egyptians constructed royal roads, and so did many other civilisations from the Middle East to China. However, in the western world at least, large-scale road-building dates back to the year 312 BC when work began on the Via Appia, the classical equivalent of the modern super-highway.

The Via Appia was the brainchild of the statesman Appius Claudius Caecus, and it

CONNECTING WEB *Today's roads are often so large that their shape can only be fully appreciated from the air. This almost sculptural interchange is in the south of France.*

connected Rome with the city of Capua about 130 miles (210 km) away. Unlike other roads of the period, which became impassable after heavy rain, this one was built for use in all weathers. It was also designed to last. Its foundations were up to 3 ft (1 m) deep, and its surface was paved with flat stones, laid so tightly – according to Appius Claudius's own specifications – that a knife could not be slid between them.

The Via Appia became the model for a system of roads that eventually spread across the entire Roman Empire. With unparalleled boldness, Roman military engineers drew straight lines between major towns and

MAN-MADE WATERWAYS

In 1327 one of the world's greatest engineering feats was finally completed. Running 1100 miles (1770 km) from Beijing to the port of Hangzhou (Hangchow), China's Da Yunhe or Grand Canal provided an inland waterway that connected the northern city with the distant valley of the Yangtze River. This mammoth undertaking was frequently interrupted by wars and catastrophic floods, which helps to explain why it took over 1000 years to be pieced together.

Most canals have one of two purposes: to supply water, or to improve transport links. The first canals were mainly water-suppliers, but China's Grand Canal was meant

primarily for the movement of produce and people. Because its route took it away from the coast, it also reduced the risk of cargoes being attacked by pirates in the Yellow Sea.

From the time of the Industrial Revolution canals became important arteries of trade. In Britain alone more than 3000 miles (4800 km) of canals were dug, allowing heavy loads to be pulled between cities by the power of a single horse.

The value of these inland waterways was eroded by the spread of railways, but by then canal-builders had set their sights elsewhere. For international shipping, two canals

PANAMA CANAL *An oil tanker negotiates the Miraflores locks near the Caribbean entrance. A single ship uses more than 50 million gallons (227 million litres) of water as it crosses the Panamanian isthmus.*

completed 45 years apart created the ultimate in global short cuts. The Suez Canal, built by the Frenchman Ferdinand de Lesseps and opened in 1869, connects the Mediterranean and Red Seas. Despite being only 100 miles (160 km) long, it reduces the sea distance between southern Europe and India by about 5000 miles (8000 km). The Panama Canal, which was designed by American engineers and opened in 1914, crosses Central America at its narrowest point, and is 51 miles (82 km) long. By sidestepping the need to circumnavigate South America, it reduces the distance from the Caribbean to the central Pacific by about 10 000 miles (16 000 km). The Suez Canal runs through low-lying ground, but the Panama Canal climbs to a natural lake 85 ft (26 m) above sea level. It has six giant locks that allow ships to make the journey over Central America's slender spine.

then built roads along them almost regardless of what lay in between. But once the Roman Empire fell into decline, road-building also became a forgotten art. The next significant step did not come until 1815, when the British engineer John McAdam started to build roads with a hard raised surface, specially curved to throw off rain. At a time when road travel often meant becoming stuck in mud, these 'macadamised' highways quickly became popular, and they were the forerunners of the roads we use today.

INTRUDERS UNDERGROUND

When any new road is built, the impact on its surroundings is immediate and obvious. Cuttings and tunnels slice through hills; embankments ride over low-lying ground; nature's lines of communication are severed and replaced with our own. But road-building affects more than the landscape in which a road stands; in order to produce smooth surfaces and gradients, modern roads use up large amounts of rock and cement, and like so many of our raw materials these have to be won from the ground.

Human interest in the Earth's buried resources goes back a long way. Our distant ancestors dug out mineral pigments such as ochre over 40 000 years ago, and also hacked their way into soft ground to extract valuable flints that could be made into tools. The Egyptians were extracting and smelting iron ore more than 3000 years ago. By 500 BC the ancient Greeks were digging out the ores of gold, silver and lead, and when the Roman Empire reached its height, lead was a particularly sought-after commodity. About 80 000 tons of this versatile metal were produced every year, and it was used not only for plumbing and making paint but – bizarrely and unhealthily – even as the major ingredient of a fashionable artificial sweetener. Like the ancient Egyptians, the Greeks and Romans also quarried stone, and in their hands it was often transformed from a mere building material into something sculpted with breathtaking skill.

There are two main ways of extracting rock or minerals from the ground. The first – usually by far the easiest – is simply to dig or cut it out. Most of the clay and limestone

needed for cement are produced by this method, and so are rocks such as marble and granite. In some places, such as Carrara in Italy's Apuanian Alps, particularly lustrous forms of marble have been quarried for centuries, leaving hillsides pockmarked by the workings.

With something as valuable as marble each giant block has to be detached with great care, to prevent accidental falls that could cause expensive cracks. But mineral ores are often crushed before they are processed, so they do not need such a delicate approach. The business of extraction can proceed at a greater rate, and even after a few decades the results can be truly extraordinary. At Bingham Canyon in Utah, for example, more than 3 billion tons of

EARTH REVEALED At its broadest, the copper mine at Bingham Canyon in Utah is about 5 miles (8 km) wide. It is large enough to generate its own local climate.

CARRARA QUARRY *The milk-white stone from this Italian hillside has been used to produce some of the world's finest sculptures.*

copper-bearing rock have been stripped away, creating the world's largest artificial hole, currently over 2600 ft (790 m) deep. But sometimes working from the surface is impossible, and then the only alternative is to tunnel underground.

As early miners discovered, the Earth does not readily admit human intruders. England's Neolithic flint miners, working in an area known as Grime's Graves, used antler picks to dig bottle-shaped holes and connecting galleries, but ventured only a short distance beneath the surface. The Romans were far more skilled at tunnelling, but their mines were still little more than pinpricks in the Earth's crust, and even medieval mines extended only a few hundred feet. The reason was simple: with increasing depth comes increasing pressure, and an ever-growing risk from collapsing rock, underground water

STOPPING THE TIDE
*This automatic flood barrier,
near the border between
Belgium and the Netherlands,
blocks tidal surges that in the
past have claimed many lives.*

and deadly gas. By the coal-hungry 19th century, mining technology had made many advances. Mines now routinely reached more than 1500 ft (460 m) underground, and spread horizontally for several miles. Nineteenth-century miners had the benefit of the safety lamp, invented in 1815, which helped them to detect pockets of potentially explosive methane seeping from the coalface, and they also had better tools for cleaving the rock. But despite these improvements mining was still a uniquely demanding and dangerous trade.

In his novel *Germinal* the French writer Emile Zola vividly described the sort of conditions that coalminers had to endure. They moved by 'dragging themselves around on their knees and elbows, unable to turn around without bruising their shoulders'. Even with pit-props holding up the immense weight overhead, the danger of a rock-fall was ever present. In Zola's chilling words, each miner was 'wedged between his two levels of rock, like a greenfly caught in the pages of a book which threatened to slam suddenly shut'.

THE SHIFTING SURFACE

In modern mines mechanisation has done much to eliminate these dangers, but, with their ever-growing scale, mines and tunnels can have a major effect on the Earth's outer skin. In most cases these effects are the anticipated results of excavation, but sometimes hidden stresses build up underground, triggering off movements at the surface. These movements often become noticeable only when they cause cracks in houses and roads, or tip trees away from the vertical. However, sometimes they take place on a bigger and more awe-inspiring scale.

One of the most spectacular examples of this occurred in the West Rand region of South Africa. Its gold mines are among the world's deepest and under constant threat of

RECLAIMED GROUND *Plastic
sheets hold the ground in place
while plants become established
on an English spoil heap. Their
roots will prevent erosion.*

flooding, and they have to be pumped constantly to allow miners to reach the gold-bearing rock. Over the years a huge volume of water has been displaced, and as a result the water table has fallen by 1000 ft (300 m), allowing layers of previously waterlogged clay to dry out. As the clay dries it shrinks, and this creates gaps in strata that were previously solid. One day in August 1964 the ground above one set of mine-workings suddenly collapsed. The result on the surface was an immense crater, nearly 200 ft (60 m) across and 80 ft (24 m) deep, which swallowed several houses. More than 20 people lost their lives.

While subsidence is unpredictable, another side effect of mining is not. Whatever is being mined usually makes up just part of the material that is excavated. Coal, for example, is sandwiched between rocks that often fall away as the seams are cut, while many ores contain large amounts of unwanted waste. All this has to be disposed of, creating the giant spoil heaps that are such a characteristic feature of mining regions. At Grime's Graves in eastern England, small

mounds – now covered by grass – still mark the places where waste chalk was dumped by the flint miners of long ago. But these knolls are completely dwarfed by the waste generated by modern forms of mining. Some of today's spoil heaps contain more than 250 million cu yd (190 million m³) of waste, a volume over 60 times as big as the Cholula pyramid. Unlike small banks of chalk, they present much more hostile conditions for any living things that try to adopt them as their home.

Compared with the ground they stand on, spoil heaps are geological anomalies. Built of material that has often been brought up from deep underground, they may contain high concentrations of chemical elements that are not normally found at the surface. Many of these elements are harmless, but some – including heavy metals such as lead and cadmium – are potentially deadly to plants. The lack of soil also makes

life difficult for plants, and the steep slopes and rapid drainage often create a punishing microclimate for any that do manage to take root. Without plants there is no food for animals, so the heap can only support small predators such as spiders and mites, which survive by hunting other creatures that arrive accidentally from elsewhere.

At one time these artificial hills were simply left where they had built up, with little being done to nurse them into life. But as mining industries become more conscious of their environmental impact, some spoil heaps are starting out on a path that will eventually turn them into more benign environments. First their steep slopes are smoothed away, and pockets of soil are brought in to give plants a foothold in their bare surroundings. Trees now grow on some of the oldest of these giant heaps, but their conversion into healthy, fully functioning habitats is still a long way off. For most, several centuries will have to pass before they begin to look like natural features of the landscape.

ROLLING BACK THE SEA

No landscape has seen more change than the frontier between land and sea. Coasts are always on the move – usually slowly, but sometimes with devastating swiftness – as the sea eats into the land in one place and builds up silt and shingle in another. On coasts that are being eroded, walls can hold the sea in check, but some shoreline structures serve a very different purpose. They challenge the sea's supremacy and protect new land that has risen from the waves.

One country – the Netherlands – has come to symbolise this offensive against the sea. In prehistoric times farmers on its low-lying coast built small mounds, called *terpen*, which provided safe places to build

SURFACE SCARS *Quarrying and open-cast mining create long-lived scars on the Earth's surface. This quarry in Hawaii was originally covered by lush tropical vegetation.*

INVADING THE EARTH

The thickness of the Earth's outer crust varies from about 6 miles (10 km), beneath the ocean floor, to between 12 and 35 miles (19 and 56 km) beneath the continents. The deepest mines currently reach about 2¼ miles (3.6 km), while the deepest test boreholes drilled on land have reached 9 miles (14 km). At these depths the temperature of the surrounding rock can be above 200°C (390°F).

houses and refuges for farm animals. Some of these mounds were linked by raised banks or dikes, and from about AD 1200 onwards the dikes became extensive enough to impound tidal mud flats. Little by little the impounded areas – known as polders – were transformed into highly fertile fields, and as the centuries went by a new and level landscape was created.

When previously submerged ground dries out, it often shrinks. This is particularly true when the ground contains a lot of organic matter, because this slowly decomposes when it is exposed to the air. In the Netherlands, ground shrinkage often meant that the newly won polders were below the level of even a modest tide, and this brought a constant threat of flooding. However, the flat Dutch landscape offered its own answer to this problem. By the 17th century, battalions of windmills had been built to harness

PURELY FOR PLEASURE

What do you do if you own a great country house or palace but are bored with the landscape around it? The answer, for some landowners in 18th-century England, was to have it changed. To bring this about they called on the services of a remarkable landscape gardener, Lancelot or 'Capability' Brown.

Capability Brown's nickname was well earned. Working long before the days of bulldozers and mechanical diggers, he redesigned parks and gardens and swept away the formal and geometrical features that had long been fashionable. In their place he created landscapes that tried to capture the very essence of nature, with graceful grassy slopes, gurgling streams, tranquil lakes, and views of peaceful woodlands silhouetted against distant horizons.

These extraordinary transformations required a huge amount of labour. Millions of tons of soil and rock were shifted by hand to build up beguiling contours, and lakes were created either by damming streams or by excavating holes in the ground. To waterproof a lake-bed Brown's labourers used a chalky clay called marl. The marl was scattered on the ground, wetted, and then hammered flat.

Many of Brown's landscapes survive to this day at such stately homes as Blenheim Palace in Oxfordshire and Alnwick Castle in Northumberland. More than two centuries after they were completed, they show how fashion alone can transform the Earth's surface.

ART AS NATURE *This artificial lake at Blenheim Palace was created by Capability Brown.*

Under natural conditions, many rivers burst their banks during times of heavy rain, and spill out over wide plains that shoulder the surplus water. In some parts of the world – such as the Mekong River delta in South-east Asia – seasonal floods are an accepted part of the annual calendar. Houses are built on stilts to keep them dry, and farmers welcome the fertile sediment that the flood-waters bring. Even so, the threat of being flooded is never a comfortable one, and to tackle it people have frequently imitated nature in a bid to keep rivers in their places.

Every year large rivers carry huge volumes of suspended matter downstream. The Mississippi, for example, transports about 300 billion tons of mud and rock scoured from the heartland of North America, while the Huang He or Yellow River sweeps along a record-breaking 2000 billion tons of silt from central China. As long as the water flows briskly, most of this sediment stays on the move until it nears the sea, but when a large river bursts its banks the current suddenly slackens. Like someone abandoning unwanted luggage, the river promptly drops its sediment, and when the flood subsides it reveals a landscape that is cloaked by silt.

Much of this silt is spread out over the ground, but near the river itself it often forms large banks called levees. Levees usually run parallel to the water's flow, and they provide a tempting foundation for confining a river to its bed. They have been used for just this purpose in many parts of the world, and over many generations have been built up into giant ramparts flanking many miles of river. In North America the Mississippi levees, which were begun in 1717, are now the largest of their kind, and extend for over 1700 miles (2730 km) along the Mississippi itself, with an even greater length along its tributaries. This makes

the dependable sea breeze, and these pumped unwanted water into canals that led away to the sea.

Since those times technology has moved on, but the principles of land reclamation remain much the same. The windmills that once graced the Dutch landscape have been replaced by less elegant pumps, powered initially by steam but now by diesel engines and electricity; the hand-operated sluices that prevent the sea from flowing back up drainage canals have been replaced by ones operated at the touch of a button. Giant flood gates, controlled by computer, stand ready at strategic points along the main shipping canals, ready to swing into place in the event of a dangerously high tide. The newest polders are protected by the longest sea barrage in the world, and altogether two-fifths of the country now consists of man-made land.

Over the centuries Dutch engineers became so skilled at these drainage operations that their expertise was in demand all over Europe. Between 1634 and 1652 Cornelius Vermuyden organised the drainage of the Cambridgeshire fens in eastern England, and other Dutchmen worked as far afield as Italy and Russia. However, the sea has also been forced into retreat in other parts of the world, sometimes for much longer. In the Xi Jian or Pearl River delta in China, near the ancient city of Guangzhou (Canton), land reclamation first began about 6000 years ago and has now created nearly 600 sq miles (1550 km²) of land – an area three times the size of Singapore.

FIGHTING FLOODS

One of the strangest sights in reclaimed land is that of boats and ships threading their way along waterways perched high above the surrounding fields. This is no contravention of the usual laws of physics, because the flowing water still makes its way downhill, albeit at a gentle gradient. However, this kind of water has a dangerous potential. Given the slightest opportunity it will quickly run towards the lower ground, and what begins as a harmless trickle can turn into a catastrophic flood.

them among the most massive engineering projects man has ever undertaken.

For most of the time these earthworks are very successful in containing floodwaters. There is, however, a catch. Because they prevent a river from spreading outwards during times of flood, levees exaggerate the water's rise and fall, making high banks an absolute necessity. If changes occur upstream – for example, if farming creates an increase in sediment – the riverbed builds up, and the

BREAKING THE BANKS *In July and August 1993 parts of the Mississippi River reached levels nearly 50 ft (15 m) above normal. More than 20 million acres (8 million ha) of land were flooded.*

river and its banks climb higher and higher above the ground. The result is a 'suspended river', created not by the ground being lowered, as occurs when land is drained, but by the water itself being lifted up.

Eventually the vicious circle has to break. The river finds a point of weakness, and breaches its man-made walls like blood pouring out of an artery. Once this kind of flow has begun, the force behind it is so great that it can prove very difficult to staunch. The Mississippi has often burst its man-made banks – most recently in 1993. In October 1887 the Huang He flooded over 50 000 sq miles (130 000 km²) of farmland, and during this catastrophe – the worst in history – nearly a million people lost their lives.

Disasters like this have forced experts to rethink the way large rivers are controlled.

Today the Mississippi is still kept in check by its levees, but other measures are also used to calm its dangerous surges. These include areas of land set aside as 'holding bays' for floodwater, and straight channels cut in places where the river formerly meandered across its plain. Channelling the river speeds up its flow and helps to make sure that water moves on before it has a chance to do any damage. A curious side effect of this is that the Mississippi, while still North America's longest river, is now over 150 miles (240km) shorter than it was before.

STOPPING THE FLOW

When floodwaters pour down a major river, sediment is not the only thing they carry away. They take water that could be useful in times of drought, and also an intangible but equally valuable commodity – the huge reserves of energy that are dissipated when water flows downhill to the sea.

Some of the first dams and reservoirs, built in ancient Egypt about 5000 years ago, functioned solely as stores, holding back a little of the precious water brought down by the Nile's annual flood. In terms of volume, these artificial pools were tiny and quite insignificant compared with the amount of

JOURNEY'S END *Stretching into the Gulf of Mexico like an outstretched hand, the Mississippi delta builds an ever-changing landscape of silt.*

HIGHER RISE *When it was
built in the 1930s the 700 ft
(210 m) high Hoover Dam was
one of the tallest in the world.*

water flowing in the nearby river. Modern dams are many orders of magnitude larger: the Aswan Dam, for example, completed in 1964, straddles the whole of the Nile. It is over 2 miles (3.2 km) wide, and holds back a lake more than 250 miles (400 km) long which is clearly visible from space.

Dams built on this scale are new features on the Earth's surface, and their full effects are only now becoming clear. On the positive side, dams undoubtedly help to supply water and regulate water flow. In dry climates large amounts of water can evaporate from a reservoir's surface, but the huge water bank held behind a great dam is still an immensely valuable resource. Dams also provide large

amounts of electricity, and they can be controlled at the touch of a button. If demand suddenly rises, the turbine gates are opened, water pours through them, and extra power is almost instantly available for use.

With these advantages it is not surprising that giant dams have been among the 20th century's greatest construction projects. There are now more than 100 dams over 500 ft (150 m) high, and most of the world's major rivers, except those that flow into the Arctic, have been dammed, often in several places. Giant dams are still being built in South America and Asia. One – the projected Three Gorges Dam across China's Yangtze River, expected to take 20 years to build – will be the world's biggest, with a generating capacity eight times that of Aswan.

However, the great dams of the future will face much more controversy than those of the past, mainly because of the likelihood of unwanted side effects. Apart from drowning useful land, dams have far-reaching consequences for the life around them. They prevent fish from migrating and create habitats for waterborne pests, and they make

areas lower down less fertile by trapping the sediment that otherwise washes downstream. Their environmental credentials are also marred by the fact that building them requires an immense amount of energy.

Sediment – or lack of it – lies at the heart of other problems that dams can trigger off. When water enters a reservoir and slows down, it sheds its sediment just as readily as floodwater spilling over a plain. If the river's sediment level is low, it builds up only slowly on the reservoir bed. However, if it is high, it can accumulate at a rapid rate and limit the life of a dam. For most dams, sedimentation will take decades or even centuries to become a serious problem, but for the San Men Xia Dam on China's Yellow River it intervened much more quickly. Its reservoir filled with silt in just four years, after which the dam had to be torn down and rebuilt.

The final problem facing the designers of large dams is perhaps the most unexpected, and potentially the most hazardous. As millions of gallons of water build up behind a dam, its sheer weight presses down on the Earth's surface, squeezing the rock beneath. More importantly, water is also driven into the rock, infiltrating its pores and cracks and subtly altering its physical characteristics. With the exception of nuclear explosions, the forces involved are the greatest that mankind has ever unleashed, and saddled with this burden the Earth responds.

After the completion of many large dams, including the Kariba Dam in Africa and the Hoover Dam in the USA, earthquakes have shaken the ground. Six have had magnitudes of over 5 on the Richter Scale, and several have been powerful enough to trigger waves that have crossed and recrossed the filling reservoirs. In most cases the aftershocks have slowly died away, and the Earth has resumed its silence, but its shaking carries an ominous message. With dams as with other structures, there are limits to how far we can go in reshaping the land.

OUR GLOBAL IMPACT

3

TRAPPED *As the human world expands, animals like the tiger face a struggle to survive.*

HUMAN INFLUENCE HAS SPREAD ACROSS THE EARTH, AND TODAY FEW PLACES ARE BEYOND THE REACH OF MAN-MADE CHANGE. MODERN TECHNOLOGY CREATES MATERIALS THAT PUSH BACK THE LIMITS OF OUR PLANET'S NATURAL CHEMISTRY; IT ALSO PRODUCES EVER FASTER MEANS OF TRANSPORT THAT MAKE THE EARTH A SMALLER PLACE. AS WE PURSUE THESE TECHNOLOGIES, NEW POSSIBILITIES OPEN UP FOR US, BUT NEW DANGERS SOMETIMES COME IN THEIR WAKE. AMONG THESE ARE THE THREATS OF CLIMATE CHANGE AND OF IRREVERSIBLE DAMAGE TO THE NATURAL WORLD. THESE CHANGES HAVE FORCED US TO RETHINK OUR PRIORITIES: HOW TO BENEFIT FROM THE ADVANCE OF TECHNOLOGY WHILE MINIMISING ITS ADVERSE IMPACT.

ARCTIC RADIATION *A scientist tests Arctic ozone levels.*

INDUSTRIAL EARTH

Life without modern industry would be hard to imagine.

Industry generates work, wealth and many products that improve our lives – it also produces pollution. Its future task lies in generating the technology needed for a cleaner world.

The world is full of places that can claim to have had a lasting impact on human history, and on the way we now live. Some are the scenes of great conflicts or of natural disasters. Others are places where important leaders came to power, or the sites of crucial scientific breakthroughs. However, of all of them, one in particular can claim to have had the most far-reaching effects on our lives today. That place is surprisingly little known.

Flanked by the leafy countryside of the English Midlands, the small town of Coalbrookdale perches above the picturesque valley of the River Severn, where steep wooded slopes drop down to the water below. Scattered among the trees is a collection of old brick buildings whose chimneys now gape blankly towards the sky. Two hundred years ago, however, they belched smoke and soot, for despite its tranquil appearance this region was once a scene of unprecedented activity. What happened here helped to trigger a

COALBROOKDALE BY NIGHT
*Swathed in crimson smoke
(opposite), these furnaces in
the English Midlands, painted
by Philip de Loutherbourg,
formed one of the world's first
industrial landscapes. Another
scene from the same region
(right) was painted by the
English artist Paul Sandby in
1803. Although haphazard to
modern eyes, these early
ironworks were a wonder of
their time.*

revolution that is still reshaping the world.

As the 18th century drew to its close, the German-born artist Philip de Loutherbourg visited this corner of England and recorded his impressions of what he saw. In a canvas called *Coalbrookdale by Night*, he depicted the brick buildings thrown into black silhouettes by a lurid crimson glow, with flaring sparks flying up into the darkness. The unearthly light reveals a collection of heavy metal pipes designed for carrying steam, while in the foreground a pair of horses – frightened perhaps by the noise and flames – drags a railway wagon towards some unknown destination. It looks like an infernal vision, but what de Loutherbourg painted was something much less sinister: an ordinary night's ironmaking in one of the world's first industrial towns.

Unlike great battles or scientific breakthroughs, the rise of modern industry cannot be pinned down to a particular date, or even to a particular invention. Nevertheless, there are some important milestones along the way. Historians generally agree, for example, that the Industrial Revolution began with the mechanisation of textile mills in northern England and Scotland in the late 17th century, when they started to use water mills or steam engines as a source of power. Later the improved ironmaking techniques pioneered in Coalbrookdale gave mechanisation a tremendous impetus, and from then onwards each new development seemed to

spawn the next. By the early 19th century, centres of production were drawing people in from the surrounding countryside, and with the help of this new urban labour force the world's earliest industrial areas came into being.

Since that time, industry has spread across much of the globe, and its products have become an integral part of modern life. Industry creates work and wealth, but it can also create great environmental change. A hint of that change was already visible when de Loutherbourg, sketchbook in hand, ventured down the slopes of the Severn valley and into the crucible of the Industrial Age.

CATALYST OF CHANGE

The hallmark of any form of industry – whether it makes metals, medicines or microelectronics – is specialisation. People and raw materials are brought together in one place, so that products can be manufactured on a large scale. The word 'manufacture' comes from the Latin *manus*, 'hand', and *facere*, 'to make', and traditional manufacturing involves just that: making things by hand. By contrast, industrial manufacturing uses machines, and its products are often destined for sale far away from the places where they are made.

Industrial development, like the Industrial Revolution itself, acts as its own catalyst. Once one industry has become established in a particular place, others frequently follow. In parts of the world where this has happened, the natural environment is often changed beyond recognition. For example, as long ago as 1835, Manchester – a rapidly growing industrial city in northern England – shocked the French historian Alexis de Tocqueville, who came from a country where industrial development was only just starting. He was impressed by the size of the factories, but dismayed by the noise and smoke, and by the 'fetid, muddy waters, stained with a thousand colours by the factories they pass'. He noticed another important side effect of industrialisation: the surrounding countryside had fallen into decay, so that it formed an uncomfortable transition zone between the burgeoning city and the rural land beyond.

In the following 100 years this kind of transformation affected many regions across the Northern Hemisphere, where 'smoke-stack industries', such as steelmaking and engineering, became established. Separating a metal from its ore requires a large amount of heat, and in most cases this comes from burning fuels. Already in the 18th century furnaces were damaging the environment

AERIAL OVERLOAD *In the former USSR a processing plant pours smoke into the sky. Scenes like this were once typical of heavy industry worldwide.*

by releasing toxic gases in smoke – in Sweden, for example, the botanist Linnaeus noticed how one large copper smelter had blighted the surrounding vegetation with its fumes. Scaled up hundreds of times to its modern equivalent, this form of industrial pollution can affect a wide area, killing trees and water life by creating acidic rain. Heavy industries also use large amounts of water during their production processes, and if it is left untreated this too can have some damaging effects.

In most modern industrial cities the smokestack days are past, but there are places where old technologies linger. One is the Russian city of Noril'sk, among the remotest industrial centres on the Earth. Noril'sk lies in the Arctic about 1700 miles (2700 km) east of Moscow, on the northern edge of Siberia's great coniferous forest. The native peoples of this region traditionally live by hunting reindeer and Arctic

foxes and by fishing, but because they are thinly spread their impact on their surroundings is minimal.

The 300 000 people of Noril'sk live in a very different way. Their city leads the world in the production of several important metals, including nickel and cobalt, as well as being a leading supplier of coal. The skyline is dominated by giant chimneys

pouring smoke from furnaces and smelters, together with vast clouds of invisible sulphur dioxide. Noril'sk is thought to be the world's largest single source of this toxic gas, which may help to explain why the forest has died for miles around. The smoke is so dense that it often creates a murky haze over the shores of the Arctic Ocean, more than 250 miles (400 km) away. For many decades

The significance of Wöhler's achievement is hard to overstate, because it finally shattered one of the fundamental scientific principles of the time. In the early 19th century, chemists divided substances into two quite different groups: organic chemicals, which contained carbon and were found in living things, and inorganic ones, which did not contain carbon, and were found only in non-living matter. The two were held to be quite separate, like parallel universes that intertwined but never actually connected. Wöhler's synthesis dealt a crippling blow to this idea. By making urea – an organic substance – from inorganic ingredients, he showed that organic chemicals could be created on a laboratory bench.

THE ORGANIC REVOLUTION

The discovery generated immense interest in this branch of science, and chemists in a number of countries quickly began to appreciate the almost limitless possibilities it opened up. Carbon is unique among the elements in its ability to form complex compounds with other substances, and many hundreds of carbon-containing compounds are found in living things. The world of organic chemistry also encompasses a vast number of chemicals that are theoretically possible but that have never been found in nature – or not yet. Like empty spaces in some infinite crossword puzzle, they lie waiting to be filled in.

the river P'asina, which flows north towards the Arctic, has served as a conduit for Noril'sk's liquid waste. Many of the Arctic's large mammals – including seals, walruses and whales – contain traces of industrial waste, and much of that is a product of Noril'sk and its furnaces far away.

The industries of the early 18th century mostly dealt with metals and mechanics, but in 1828 something happened that eventually led to a completely new field of industrial endeavour. Using simple chemical ingredients, the German scientist Friedrich Wöhler managed to synthesise urea, a substance produced by animals. With this breakthrough he ushered in a chemical revolution that would eventually match the mechanical one that had gone before.

During the last 150 years chemists have identified hundreds of thousands of organic compounds, and have found ways to produce them in useful amounts. Often we do not connect these substances with industry at all, but they are industrial products just as much as nuts and bolts. For most of us, life without them would be difficult to imagine, and for some people it would be impossible.

One of the most valuable was originally obtained from the bark of willow trees, until in 1859 Hermann Kolbe, a one-time pupil of Wöhler, found a way of producing it in the laboratory. Its formal name is salicylic acid, but it is much better known in its slightly modified form as aspirin. Although several decades passed before its value became appreciated, aspirin went on to become one of the world's most widely used drugs, and its list of useful properties

FINAL JOURNEY *Drums of chemical waste await incineration at a special disposal plant in southern England. High temperatures destroy the majority of organic chemicals.*

is still growing. Aspirin belongs to a family of medicines that can be produced from the most unlikely of raw materials. Paracetamol, for instance, was originally made from the tar distilled from coal – a substance that was the springboard for a whole constellation of new organic substances, ranging from brilliant dyes to perfumes and solvents.

Aspirin and paracetamol make our lives more comfortable, but other industrial products intervene in life in even more significant ways. During the same decade that Kolbe synthesised aspirin, the British chemist Alexander Parkes experimented with a recently discovered substance called pyroxylin. He found that if it was mixed with camphor, and then treated with a solvent, the result was a solid substance that became flexible when warmed. Once this flexible state was reached, it could be reshaped, and it would retain this shape when it had cooled. This curious but, at the time, apparently useless substance was the world's first synthetic plastic.

If Parkes could be brought forward to the present day, he

WHALES AS SPECKS *Seen from the air, 1 ton beluga whales look like a collection of white specks in an estuary of the Canadian Arctic. Belugas are severely affected by industrial pollution.*

would be astonished to see where his chance discovery had lead. One of the great values of plastics is that they can be moulded to fit intricate shapes with great accuracy. Another is that they are light, and a third is that, once made, most of them are chemically inert, shunning any kind of interaction with the substances around them. For manufacturers of artificial human heart valves, just as for manufacturers of contact lenses, they are substances that have no equals. As much as rapid travel and instant communications, they represent the benefits of life in an industrial age.

A GLOBAL REACH

There is, however, a less welcome side to this chemical cornucopia. At one end of the scale is the problem of durability, and of materials that refuse to break down once their useful life is over. This is a general difficulty with plastics, because microorganisms are unable to break the chemical links that hold them together. At the other end of the scale is another problem of durability – one in which individual molecules of the compounds end up in places where they may have damaging effects.

In the St Lawrence River in eastern Canada, a small population of belugas or white whales provides evidence of where this can lead. Despite being protected against hunting, and having no predators in the region, the belugas are slowly dwindling. The reason, according to ecologists who have studied them, is that they are contaminated with highly toxic organic chemicals called polychlorinated biphenyls, or PCBs, which were once used to make plastics and insulating materials. PCBs are now banned in North America and many other industrialised countries, but they continue to leak out of abandoned factories and warehouses, and into streams and rivers.

The danger from PCBs, and chemicals like them, is twofold. Because they resemble substances found in living things, they act like the molecular equivalent of Trojan horses and are absorbed by animals and accumulated in their tissues. Once stored away, they are then in a position to inflict the maximum possible damage. In belugas and other marine mammals, these on-board poisons seem to disable the immune system; they also imitate the effects of some reproductive hormones, so they impair the animals' ability to breed. The second hazard is their persistence. Most organic chemicals are soon broken down in nature, but PCBs can remain intact for decades. Dissolved in water or trapped in seabed sediments, they remain a long-term hazard for many kinds of life.

The majority of organic chemicals, including PCBs, are never meant to escape into the outside world. With substances

such as plastics and resins it is fairly easy to ensure that this does not happen. Volatile compounds, on the other hand, are a different matter.

Given the slightest opportunity, volatile chemicals evaporate and vanish into the surrounding air. On a domestic scale this characteristic may be no more than a minor irritation, most often encountered when the lid is accidentally left off a can of paint. However, scaled up to an industrial level, it has some major implications. Whenever volatile compounds are used in industrial processes, some proportion inevitably ends up in the air; if they form part of a product, their eventual escape is almost certain when that product is scrapped.

In the United States, the world's leading producer of organic chemicals, roughly

5 million tons of these volatile substances are released every year. What happens to them depends on their individual chemistry. Styrene, for example, which is used to make the plastic polystyrene, breaks down within a few hours, while benzene, which is used in a wide range of industrial processes, breaks down in a few days. Dichloromethane, a widely used solvent, lasts for up to two years, while trichloroethane, another solvent, lasts for at least six. By contrast, carbon tetrachloride lasts for over 60 years. Dichlorodifluoromethane or CFC-12, which was once widely used in aerosols, refrigerators, freezers and insulating foam, has an atmospheric lifetime of over a century.

Many of these volatile chemicals are poisonous in high concentrations, but if their life in the atmosphere is fairly short they

REPAIRING THE OZONE HOLE

Ozone is the Jekyll-and-Hyde of atmospheric gases. At ground level it is a potent poison, but in the stratosphere – between 10 and 21 miles (15 and 35 km) above the Earth's surface – it shields life on the ground from damaging ultraviolet light from the Sun. When the existence of the Antarctic ozone hole was first confirmed in 1984 the danger to living things was immediately apparent, and many of the world's countries joined in discussions aimed at combating this dangerous side effect of industrial pollution. The result was the Montreal Protocol, agreed in 1987, which aimed to phase out gases such as CFCs that play an important part in disrupting the ozone layer. Nine years later, in June 1996, atmospheric scientists revealed the first positive results from

this emergency action. For the first time on record the concentration of ozone-destroying gases fell in the lower atmosphere, providing a hopeful sign that levels in the stratosphere are likely to follow suit. This positive development is just the first step in the battle to repair the damage to the Earth's ozone

shield. According to current projections, the recovery of the ozone layer will be relatively rapid in the Northern Hemisphere, but over the Antarctic, where temperatures are much colder, the process will take longer. Here the ozone hole is not expected to be filled until the middle of the 21st century.

OZONE RESEARCH **A balloon readies for liftoff during the long Arctic night. Balloons are used to investigate the complex chemistry of the ozone layer.**

have little chance of building up to dangerous levels. However, chemicals such as carbon tetrachloride and CFC-12 present a quite different problem. Since the 1930s about 2 million tons of carbon tetrachloride have been released into the air, while in just 25 years the figure for CFC-12 and closely related compounds grew from almost nothing to more than 10 million tons. As a result, these volatile chemicals have had a chance to build up to greater concentrations than any other synthetic substances ever released.

Scientists first warned in the 1970s what the effect of this pool of organic chemicals might be, and today that effect – the thinning ozone layer – is a well-established reality. The ozone layer normally acts like a transparent, high-altitude shield, protecting living things from the Sun's harmful ultraviolet radiation, but that shield has now been so eroded by CFCs and other chemicals that its screening power has been seriously weakened. This effect is most marked near the poles, where the ozone layer is effectively holed for several months each year.

Thanks to an international agreement signed in the 1980s, and strengthened since, many ozone-destroying chemicals are now

KITCHEN CHEMISTRY *Old refrigerators are stacked up in a dump in Germany. The CFCs they contain will be removed and then chemically destroyed.*

either banned or due to be phased out in the near future. However, throughout the world industries and industrial products continue to release countless other solvents, refrigerants and pressurised propellants. Quite how each of them will affect the atmosphere's chemistry remains to be seen.

When Alexis de Tocqueville visited Manchester in the early days of the Industrial

POLLUTION ROMAN-STYLE

Analysis of ice locked up in glaciers shows that lead pollution has a long history. In Roman times, lead levels in the atmosphere were over four times the natural background figure because lead was widely mined. They then fell back, but by 1500, with mining widespread in northern Europe, they were eight times higher than they would have been without human intervention. By comparison, recent ice laid down in Greenland contains up to 140 times the natural background levels.

Revolution, industry was something that affected a very limited part of the planet, and in the first part of the 20th century, the world still seemed quite big enough to absorb the effects of industry, even if there were areas where pollution was clearly a problem. Today the chemical evidence of industry can be found almost anywhere, from the upper part of the stratosphere, about 21 miles (35 km) above the ground, to the sediment lying on the surface of the seabed. Industrial by-products are trapped by snowflakes falling on remote ice caps and by leaves in the heart of rain forests. They even affect animals that live deep in caves, where running water provides the only contact with the outside world.

GETTING RID OF WASTE

Since the beginning of the Industrial Revolution we have shown ourselves to be extraordinarily ingenious in devising new machinery, creating new

substances and manufacturing new products, but not nearly as imaginative in dealing with the side effects of this inventiveness. In the mid 19th century toxic chemicals were simply dumped onto the ground around factories or into nearby rivers; 100 years later, not much had changed. In the late 1950s residents of the town of Minamata in the southern Japanese island of Kyushu suffered devastating poisoning from eating fish contaminated with mercury by local industries, while in Yokkaichi, a city northeast of Osaka, cadmium pollution made many people ill. In 1978 part of the city of Niagara Falls in New York State was

PAINTED RIVER *Mining waste stains the waters of the Rio Tinto in southern Spain. Copper has been mined in the area since Roman times and before, and pollution, too, dates back that far. The Rio Tinto owes its name, meaning 'Tinted River', to its polluted waters.*

WATERBORNE POLLUTION *When it reaches extremes like this, water pollution is easy to spot. Other forms of pollution can be much more difficult to identify.*

declared a federal disaster area. Known as Love Canal, this harmless-sounding site was used to dispose of toxic waste, then covered over when it was full and the land used for schools and houses – until the waste started to leak out.

Despite their severe consequences, the events at Minamata, Yokkaichi and Love Canal were problems on a local scale, and could be seen as matters of local housekeeping rather than of global significance. However, the discovery of the Antarctic ozone hole in 1984 provided dramatic confirmation of our ability to change the entire world through the things we make and use. In years to come, this single event – rather like the first view of the Earth from space – will probably prove to be a crucial turning point. After more than two centuries of transforming the environment, tomorrow's industries will have to be much more careful about how they fit into the world around them.

It is often said that in nature there is no such thing as waste. Although true in its essentials, this statement glosses over an important fact. A vast amount of waste matter is generated in nature, but all of it – from carcasses and fallen leaves to feathers, fur and droppings – is rapidly used by living things. The reason for this is simple. Organic substances take energy to make, and this energy is locked up in their chemical bonds. If something can break open these bonds it can harness the energy they contain, and it can also acquire chemical building blocks to build itself up.

ZERO-EMISSION INDUSTRY

Throughout life's immensely long history on the Earth, complex networks have evolved that link each species with the others around it. These networks are food webs – pathways that connect different

heat is lost, but in Kalundborg it is used in a nearby pharmaceutical plant, and also in an oil refinery and a fish farm. The burning coal produces sulphur-charged smoke, but before the smoke escapes into the air its sulphur is removed. This gets rid of a potential pollutant, but it also creates a useful raw material. The sulphur is passed on to a processing plant where it is used to make gypsum, the main constituent of plasterboard, used in building. Plasterboard has to be heated during its production process, and this heat comes from natural gas from the refinery, which would normally be burnt off. The pharmaceutical company provides fertiliser for the fish farm, and the fish farm generates organic fertiliser that is used on local fields.

This kind of industrial web is still very unusual, and it is by no means perfect because, unlike a natural food web, it contains many 'loose ends'. However, the principle has now been taken up in several other parts of the world. In Chattanooga, Tennessee, plans are under way to build one of the world's first 'zero emission' industrial areas on a site currently occupied by derelict buildings. Using each other's by-products, companies would quite literally profit from their neighbours' waste.

One of the lessons of ecology is that the fewer species a food web contains, the more liable it is to be disrupted if one of them suddenly declines or disappears. In nature many food webs contain so many species

species with the things that they eat, and with the things that eat them. A typical food web links dozens or hundreds of species. In each web, matter and energy are both constantly on the move, and nothing accumulates as unwanted refuse. Human industries may seem far removed from this kind of natural network but, as scientists have long recognised, the worlds of ecology and economics have a lot in common. If industries can be made to work like organisms in food webs, the problem of unwanted waste could be sharply reduced.

In the coastal town of Kalundborg in Denmark, the industrial equivalent of a simple food web has been operating successfully for more than 25 years. In this town a coal-burning power station generates electricity, but like all power stations it also creates waste heat, which is given off when the steam that drives turbines is condensed back into water. Normally this that they are cushioned against changes of this kind. For an industrial web to work effectively, diversity is also essential.

Not everybody would welcome a large and varied industrial community on their doorstep, but, as in a natural food chain, the different participants would not necessarily need to be in exactly the same place. In nature, many animals travel from place to place, linking one food web with another. The same already happens in industry, with useful waste being moved so that it can be recycled. At present this kind of trade is quite limited, but in the future much larger cross-connections could be created. With a system like this, each raw material might contribute to dozens of different products, just as food made by plants passes through many different living things.

REDUCING POLLUTION

Creating the industrial equivalent of food webs will undoubtedly take a long time, and the 'zero emission' factory is likely to remain a distant goal. Until this kind of new planning becomes widespread, other measures will be needed to protect the Earth

NATURAL POLLUTION

Pollution is any form of contamination – physical or chemical – that is harmful to living things. In most cases it results from human activities, but we are not alone in polluting the planet: nature itself contributes its share. Highest on the list of natural pollutants are volcanic eruptions, which release huge quantities of toxic gases into the air along with particles of dust. A single large eruption, such as that of Krakatoa in 1883, can destroy plants and animals over a wide area and inject so much dust into the atmosphere that the amount of energy reaching the Earth's surface may fall by a tenth. Nature also causes pollution in less dramatic

BED OF IRON *The iron oxide in this Icelandic stream is natural. It has been dissolved by hot water underground and then deposited as the water cools.*

ways. The atmosphere is contaminated by dust from desert storms. Where the underlying rock is granite, radon – a radioactive gas – sometimes seeps to the surface, and in some places oil or tar find their way upwards from below. Other living things also generate some of the substances that are considered pollutants when released by human beings. For example, most of the nitrogen oxides in the air are emitted not by factory chimneys or cars but by bacteria that live in soil.

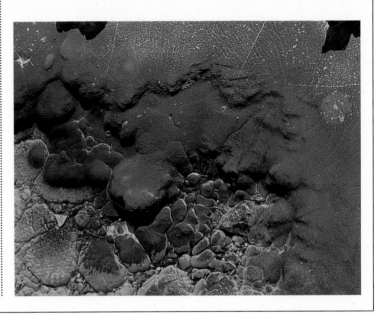

from the far-reaching effects of industrial processes and of the products we use.

In the warm waters surrounding the Florida Keys, corals provide a living record that shows how effective decisive action can be. In 1921 an American chemist, Thomas Midgley, discovered that a compound of lead could be used to prevent 'knock', a metallic clanking in petrol engines. Knock is caused by inefficient combustion, and it reduces engine power and increases wear on cylinder valves. Midgley's compound, a pleasant-smelling but highly poisonous liquid called tetraethyl lead, suppressed knock and made engines more efficient.

Despite reservations by some scientists, the new fuel additive was soon on sale throughout the USA, and within a few years the levels of lead in the atmosphere were climbing sharply. Lead particles settled on the ground, and also reached the sea off the Florida Keys. Here corals built it into their limestone skeletons, producing a yearly record of how much lead there was in the environment. Readings from the reefs show that between the years 1922 and 1980 lead levels leaped upwards until they stood at

KEEPING A RECORD *By building metals and minerals into their skeletons living corals, such as these ones off the Florida Keys, produce a permanent record of pollution levels.*

GREENER AND CLEANER
In Mexico City, green taxis help to reduce air pollution by running on unleaded fuel.

four times the original figure. Then, from the mid 1980s onwards, this upward trend was abruptly reversed. This sudden change was caused by legislation that removed lead from fuel. The lead content of gasoline dropped by half in just four years, and it was not long before the levels in corals began to follow suit.

Today about 70 000 tons of lead are still added to fuel every year, mainly in developing countries. In the USA, Europe and Japan lead-free fuels are now widely available, and cars still continue to run smoothly. How do they do this without a product that was once thought to be so essential?

The answer lies in one of the most important ways of combating the unwanted side effects of industrialisation: replacing outdated technology with systems or products that are less damaging to the natural environment and to ourselves. For many decades tetraethyl lead kept its position simply because no one saw any pressing need to replace it with anything else. In the 1970s it came under intense scrutiny, partly because there was growing evidence of its dangers to health, and partly because it disabled the catalytic converters that were being developed to combat other forms of pollution from car exhausts. Spurred on by these concerns, engineers and industrial chemists found other ways to make fuel burn efficiently, and as a result a poisonous product became redundant, even though

today's vehicle exhausts are still far from pollution-free.

This sequence of events has been repeated with a number of other compounds that have proved to have dangerous effects. These include not only CFCs but also toxic substances in paints and other products, and a number of pesticides which can linger in the environment for several years with highly damaging results. By the substitution of pesticides that break down much more quickly, the problem of toxic build-up can be averted.

MORE FOR LESS

In the last 200 years industrialisation has been like an ascending lift, with different countries getting into it at different floors. Initially the lift contained just a handful of nations, but since then it has filled rapidly, and it is still climbing steadily upwards. More and more of the Earth's surface is now devoted to extracting the raw materials that industry needs, and an even larger area is taken up by factories that process those raw materials, and by the transport systems that take finished products to the

places where they are sold. Given the speed and extent of this change, it is easy to imagine that the lift must soon approach the top of the building, where it will come to an uncomfortable halt.

Surprisingly, concerns like these are nothing new. Two centuries ago, when the Industrial Revolution was barely under way, the ironmakers of Coalbrookdale faced an acute shortage of charcoal, which they used as fuel. For centuries trees had been cut down to supply the region's small-scale furnaces, and as a result wood was often in short supply. However, in the early 18th century an ironworks owner called Abraham Darby successfully experimented with coke, made by heating coal and driving off its tar and gases. This new fuel proved to be even more effective, and in time it supplanted charcoal completely.

This pattern of unforeseeable change has been mirrored in our own times. In 1952 a United States commission appointed

COLOUR TEST *A technician takes a sample of pigment for colouring ceramics. Beautiful as well as versatile, ceramics are ancient materials with a future in modern industrial processes.*

by President Harry S Truman reported on the state of the world's industrial resources, and suggested that important minerals such as copper would soon be in short supply. In 1972 a group of scientists and economists known as the Club of Rome (where they met) produced a report entitled 'The Limits to Growth' in which they repeated these warnings. They predicted that some vital minerals would begin to run out altogether by the year 2050. At this point the lift would have reached the top floor, and further industrial growth would be impossible.

In the years since, these reports have proved to be overly pessimistic. Instead technological advances have helped geologists to locate huge and previously unknown deposits of fossil fuels, while improved efficiency has helped industrial nations to make more out of less. In the USA, for example, the amount of energy used for each unit of production fell by 26 per cent between 1973 and 1987, while in Japan it fell by 34 per cent. Although there is still widespread waste, some resources are now used much more carefully than they once were. Many are recycled in significant amounts, even though the amounts we use are often continuing to rise.

In medieval times it was widely believed that metal ores somehow grew beneath the ground, so they were practically inexhaustible. We now know that industrial resources have their limits, although few of

PIPED LIGHT *A slender fibre optic thread can do the same work as thousands of copper wires. Advances of this kind are helping to make modern life less polluting.*

those limits have yet been approached. With the introduction of new materials, such as ceramics and fibre optics, different resources are being opened up which will replace ones used before. However, while we have yet to run out of raw materials, we are running out of substances that can be safely thrown away. More than anything, tackling that problem will help to shape the future of the industrial Earth.

THE EVER-SHRINKING WORLD

Until humans gained ascendancy, Nature imposed her own boundaries. Distance kept living things in their place, and seas and mountains created additional barriers. Today, as humans crisscross the globe, boundaries are becoming blurred.

On a busy morning at an airport somewhere in southern Europe, a plane arrives on a nonstop flight from South America. The journey has been long, and most of the passengers are weary. One, however, is fully alert. Sensing the wave of fresh air blowing in through the open doors, it makes its way past the people gathering their belongings, emerges into the access walkway and carries on into the terminal building. For several minutes its course seems haphazard, but then another current of air – more intermittent than the first – guides it towards an exit of a different kind. A security camera registers two passengers walking through the door onto the balcony, but fails to spot their tiny companion – a tropical sap-sucking bug that speeds over their heads and across the runway before vanishing beyond the perimeter fence.

There, in almost every case, the story is likely to end. At large in an alien world, the new arrival will probably be unable to find the plants that it needs to live and, after a few days or weeks of struggling to survive, is almost certain to die. But suppose events take a different turn. This time the newcomer not only discovers an abundant source of food but also finds that few predators stand in its way. Its body is swollen with eggs, and there are ample places in which to lay them. Within a short time its numbers have risen to thousands, and a few weeks later they have grown to millions. A major outbreak of crop devastation is under way.

This incident, although imaginary, is based on fact. Throughout history, people have travelled from one part of the world to another, accidentally or deliberately taking other forms of life with them. As methods of travel have improved, our planet has effectively shrunk. The result is that few parts of the Earth are truly isolated, and most places are easily accessible to people and to the living things they carry with them. This crisscrossing of the globe has already produced huge changes in landscapes and their wildlife, and, as travel becomes even more rapid, some of those changes look certain to increase.

THE FELLOW-TRAVELLERS

Of all the world's wild places, remote islands have been most changed by domesticated animals and nature's stowaways. Many of these animals simply escaped, but in the days of sail some – including goats and pigs – were deliberately released in the hope that they would multiply and become an easy source of food. On many islands the newly freed animals failed to find enough to eat and soon died out; on others they flourished, and instead of being hapless castaways they became the dominant forms of animal life in their new homes. Once they had built up to sufficient numbers, they were capable of altering their surroundings beyond recognition.

The island of St Helena in the South Atlantic, once the lonely place of exile of the French emperor Napoleon I, provides a particularly well documented example of how this process took place. In 1502 Portuguese sailors spotted its central peak, and managed to make a landfall on its forbidding and wave-battered coast. At some point between that year and 1513 the first domesticated animals were brought ashore. Herds of goats soon fanned out over the island's richly forested volcanic slopes, and browsed on any leaves within reach, while pigs rooted for food beneath the trees. Female goats can produce one or two kids every year, while sows can produce litters of

POINT OF DEPARTURE *Air travel brings distant places within easy reach. Despite precautions, nature's stowaways can join human travellers.*

LONELIEST ISLE *St Helena's bare mountainsides show the impact of introduced animals. Left: Over 20 species of flightless moa once thrived in New Zealand. They disappeared once the islands were discovered.*

up to a dozen piglets. Given a plentiful supply of food, St Helena's goat and pig population grew much faster than the human one and rapidly spiralled out of control.

By the time the biologist Charles Darwin visited the island a little over three centuries later, St Helena's 50 sq miles (130 km²) of once-verdant forest had undergone a biological transformation. The original tree cover, which had evolved in the absence of any large mammals, had no defences against the foreign invaders. Saplings were killed before their leaves had a chance to grow beyond reach, which meant that the forest consisted increasingly of ageing trees, without any younger ones to replace them. To make matters worse, many trees were cut down to make charcoal. In 1731 – just over 100 years before Darwin arrived – an order was issued that all the stray goats and pigs were to be destroyed, but, as Darwin himself noted, the damage was already done and was far too extensive to reverse.

During his walks across the island's interior, Darwin assessed the extent of this ecological change. He noticed that remnants of the island's original plant cover could still be seen in places that were too inaccessible for either goats or humans, but elsewhere there were few signs of St Helena's original flora. At the time of his visit about 750 species of plant had been counted on the island, but only 50 of these were native. The rest had been imported – mainly by British colonists – giving the island a strangely familiar feel to someone who was British himself. Plantations of Scots pine were scattered over the hills, and gorse, willows and brambles were everywhere. In a little over three centuries St Helena had lost not only most of its forest but also many small animals that depended on it for survival.

Changes like these have occurred on islands both large and small on many occasions in human history, and they have often been amplified by the work of settlers. For the native wildlife of these islands – such as the snails and insects of St Helena, the flightless birds in New Zealand, and the marsupials in Australia – the arrival of alien competitors, and also of efficient predators such as cats and rats, brought tremendous pressures in the struggle for survival, and resulted in many extinctions. In some cases the release of outside animals affected geology as well as wildlife. By stripping the vegetation of St Helena, goats and pigs exposed the crumbling volcanic soil to wind and rain and increased the natural rate of erosion. Today domestic animals are kept firmly behind fences, and the island is being reforested, but no amount of work will ever return St Helena to its virgin state.

PATTERNS OF CHANGE

Unlike goats, pigs and other domestic animals, many of the stowaways that accompany human travellers are too small to attract immediate attention. Hidden away among clothes or in cargoes, many have successfully used human help to cross barriers that were formerly insuperable obstacles.

The Atlantic Ocean is one such barrier. Despite the presence of scattered islands such as St Helena and the Azores farther north, which look as if they could act as stepping stones, the distances between successive steps are far too great for land-based animals to cross. Every year some migratory North American birds are blown across to Europe by the prevailing westerly winds, and occasionally North American butterflies manage to make the crossing; the striking orange-and-black monarch (*Danaus plexippus*), for example, reached the Azores in 1860 and has recently established a toehold in Spain. However, movement in the other direction – even for animals that have wings – is relatively unusual, unless they have unwitting assistance in making the crossing.

One small animal that has done this with serious results is the Hessian fly (*Mayetiola destructor*), which is fairly widespread in northern Europe. This inconspicuous insect, measuring less than $1/5$ in (5 mm) long, is a species of midge that attacks wheat. It stunts the growth of

TRANSATLANTIC TRAVELLER
The monarch butterfly has spread from the New World to the Old, settling in southwestern Europe.

TRAVELLING PLANTS

Animals are not the only forms of life that have been carried from one place to another during human exploration of the globe. Many plants have been deliberately introduced into new regions as crops, but hundreds of others have escaped into the wild, and some are now serious weeds. One of the most beautiful examples of these runaway plants is the water hyacinth (*Eichhornia crassipes*), which originally comes from South America. It was exhibited at a flower show in the American state of Louisiana in 1864 and managed to find its way into the region's waterways, where it has since become a major problem. The same plant has also clogged lakes in many other parts of the tropics, from

Africa to Indonesia. On land these stowaway weeds include dandelions (*Taraxacum*) and the ribwort plantain (*Plantago lanceolata*), both European plants that have followed colonists to North America and many other parts of the world. The common evening primrose (*Oenothera biennis*), which is pollinated by moths and flowers at night, has successfully travelled in the other direction, spreading

throughout Europe from its homeland in North America. One of the most unusual plant escape stories is that of the Oxford ragwort (*Senecio squalidus*), which despite its name actually comes from southern Germany. This plant escaped from a botanical garden in the English city of Oxford during the 19th century, and its seeds were spread across the country by train.

FLOATING BLOOMS *A carpet of water hyacinths clogs a waterway in Louisiana. By screening the sunlight it harms the wildlife of ponds and lakes.*

wheat plants, and when entire fields are infested the grain yield drops sharply.

This unwelcome traveller is thought to have reached North America in the 18th century with the accidental help of mercenaries from the German state of Hesse, who fought alongside British soldiers in the American War of Independence. The Hessians brought wheat straw as bedding for their horses, but when they eventually departed the fly remained behind. It soon spread to the wheat fields in its new home, and in the two centuries since then has become a significant pest. In the autumn adult Hessian flies lay eggs on young plants of winter wheat, and their grubs spend the coldest months in a dormant state before coming to life in the spring. In order to control the fly the planting of winter wheat often has to be delayed so that the adult flies – which live for just a few days – have no young plants on which to lay their eggs.

This European insect has also spread to parts of Africa and Australia, but Europe is

not the only continent that has exported its less desirable forms of animal life. One handsome but infamous North American insect, the Colorado potato beetle (*Leptinotarsa decemlineata*), originally lived in the Rocky Mountains, where it fed on nightshades (*Solanum*), which are wild relatives of the potato. When potato cultivation began in the Midwest in the 1850s, a vast sea of nutritious leaves and tubers opened up, and the beetle was quick to make use of this new source of food. Imported Colorado beetles were discovered in Germany in the late 1800s, and they spread across the whole of Europe, reaching Russia in

VORACIOUS JAWS *Fields of potatoes create a perfect habitat for Colorado beetles, helping them to spread far beyond their original home.*

1959 and Turkey in 1976. In North America the beetle is attacked by a parasitic fly and several kinds of predatory bugs, but in Europe it has few enemies. As a result it is more of a pest in its new home than in its original one.

SHRINKING SEAS

The Colorado beetle and Hessian fly both travelled by boat, as did the goats and pigs that disembarked on St Helena, and the cats and rats that have ravaged island wildlife in many other parts of the world. Until the age of steam, travel by sea was a slow and exacting business. Domestic animals were tended by the crew, who had a direct interest in their welfare, but cats and rats had to fend for themselves. Many other stowaways were less well adapted to life aboard ship and would have perished during the weeks or months away from land.

For European naturalists seeking to bring home animals and plants from distant parts of the world, long voyages raised many problems. Some animals, including

the famous dodo (*Raphus cucullatus*) – a flightless bird from Mauritius – managed to survive the journey back to Europe, although in the case of the dodo it seems that the few birds that did arrive alive put on pounds of extra fat during their long confinement below decks. Many other animals were transported in the form of preserved skins, which were then stuffed and displayed once they had reached their destination. Seeds were easy to carry, but living plants were much more difficult. They needed light, and this meant exposing them to the wind and salt spray, which made a deadly combination. The answer to this difficulty, a kind of portable miniature greenhouse, is generally credited to a London doctor called Nathaniel Ward. His 'Wardian cases', invented in about 1829, allowed delicate plants to endure long sea voyages without too many ill effects.

Ward's invention came at a time when the first steamships were beginning to ply the seas. At first steam was simply an adjunct to sail, but by the late 1830s ships were crossing the Atlantic under steam power alone. The new technology cut journey times considerably and released ships from the vagaries of the wind. This allowed major growth in international trade, and opened up a large increase in passenger traffic between one continent and another.

One important effect of steamships was to extend the chains of consumption that connect people with their food. In London during the 1830s, for example, the average distance travelled by imported food was about 1800 miles (2900 km). By the 1850s

END OF THE LINE *Pigs introduced onto the island of Mauritius destroyed the dodos' eggs and brought about their extinction.*

this had risen to 3600 miles (5800 km), while in the early 1900s it was nearly 6000 miles (9650 km). Grain could be carried for long distances by sailing ships, but perishable food was a much more difficult commodity to transport. The invention of practical refrigeration in the 1870s helped to solve that problem and made it possible for steamships to carry meat halfway around the world. As a result a family sitting down to eat in London could dine on meat that had been raised on farms in Patagonia or Australia.

PROTECTED PLANTS *Victorian enthusiasts used glass cases in which to grow exotic plants. Ferns and orchids were their particular favourites.*

Modern humans, for the most part, are sedentary animals. Although we travel to work we have fixed homes, and we return to them on a regular basis. In some parts of the world a few peoples do still follow nomadic pastoral lifestyles, and there are some who carry out the ancient practice of trans-humance, in which animals are driven to summer and winter pastures on a yearly cycle, with their owners moving too. However, in our increasingly urban world, peoples like these are very much in the minority.

INTO THE AIR

Despite this settled existence, the rest of us have not entirely shaken off the legacy of our shifting past. Every year huge numbers of people display a sudden behavioural change that bears similarities to patterns of behaviour shown by migrating animals. After a period of preparation the normal daily routine is suspended, and the individual concerned disappears from his or her habitat. Some time later he or she returns, and the routine is resumed. This process is called taking a holiday, and it has had a profound effect on many parts of our planet.

Until the advent of air travel most tourism was a localised phenomenon, and

the distances travelled were relatively small. Coastal areas within reach of large cities bore the brunt of the yearly migration, and in resorts from the French Riviera to New Jersey hotels sprang up to accommodate the annual influx of visitors. But with the introduction of aircraft much longer journeys became possible, and when passenger jets replaced propeller-driven planes the speed and range of aircraft more than doubled, and the scope for travel widened further. Today more than 400 million people travel abroad every year, a figure that is equal to about one-fifteenth of the world's population.

For coastal resorts the impact of visitors on this scale can completely reshape the original landscape. On the island of Cancún, for

example, off Mexico's Yucatan Peninsula, a 15 mile (24 km) stretch of almost deserted beach has been transformed into one of the world's leading tourist destinations, attracting about 1 million people a year. The lukewarm breakers of the Caribbean still tumble over the white sand, but above the high-tide mark the island's original scrub, which used to be the haunt of iguanas, hermit crabs and roosting frigate birds, has been replaced by a different kind of environment. There are carefully tended lawns studded with elegant palms, paths that lead down to swimming pools provided for those who do not care to venture into the sea, rows of guest houses and multistorey hotels. Some of the island's natural inhabitants remain, but following this sweeping change many have retreated to more tranquil surroundings.

Cancún is one of the largest developments of its kind in the world, but in many other places, from Hawaii and Florida to Spain and the Far East, long stretches of coast have become built up in a similar way. Hotels, restaurants and roads are the most obvious results of this process of change, but when coastlines are developed they also

ON THE MOVE *A caravan stops for water on the fringes of the Sahara. Trade was once the main reason for travel; today tourism is equally important.*

OCEAN PARADISE *In the Maldive Islands tourism has been carefully controlled, and the white coral beaches have kept their unspoilt beauty.*

alter in ways that are harder to spot. Instead of taking place on land, some of the most important changes occur under water.

One of the attractions of tropical seas is their vivid colour and sparkling clarity. This is partly produced by the strong sunlight, and partly a result of their physical and chemical nature. Unlike many shallow seas in temperate regions, which are frequently rich in dissolved oxygen and natural nutrients, tropical waters are often the opposite. They are the marine equivalent of deserts, and contain relatively little that animals can use as food. In these conditions some animals have evolved a remarkable way to overcome the lack of sustenance: their bodies harbour microscopic single-celled algae, which make their own food in the same way as plants, by harnessing sunlight. In return for providing their host with a share of this food, the algae gain a secure place in which to live.

The list of animal species that harbour these invisible internal partners is a surprisingly long one, and includes sponges, jellyfish, flatworms and giant clams. But of all the animals that live in this way the most important by far are the reef-building corals. A single coral may contain more

than 10 million algae, and the coral's privileged position close to the sea's surface ensures that the algae receive all the light that they need to thrive.

Coral reefs build up extremely slowly, and the individual coral animals that form them depend on stable conditions for their survival. Some coral reefs have experienced human visitors for millennia, but the presence of scattered hunters picking over their immense bulk for fish or edible molluscs has done them little harm, other than the occasional breakage of their sometimes

URBAN SHORE *Looking like jagged teeth, hotels and apartment blocks catch the setting sun. The original shore has disappeared.*

fragile skeletons. However, when humans arrive in far greater numbers, the stage is set for intervention of a more serious kind.

Construction often generates sediment, which gravity feeds into the sea, while hotel bathrooms generate waterborne waste that flows in the same direction. Both can spell

land lay beyond this crystalline barrier, but it was not until 1820 that an American sealer, Nathaniel Palmer, sighted land to the far south of Tierra del Fuego. In the same year land was seen by several other seafarers, including the Russian explorer Thaddeus von Bellingshausen, and the continent's existence, which had long been suspected, was finally proved.

Since Antarctica's discovery, the story of human endeavour in and around this

---

CIRCLING THE EARTH

The first successful circumnavigation of the Earth was led by the Portuguese navigator Ferdinand Magellan, who set sail from the Spanish port of Seville on August 10, 1519, with a fleet of five ships. Magellan himself fell ill and died in the Philippines in 1521, but his ship arrived back in Spain on September 6, 1522, after a voyage that lasted just over three years. Today satellites often circle the Earth several times in the course of a single day.

---

ultimate continent has not been an entirely creditable one. Alongside many feats of exploration have been acts of unbridled exploitation, as sealers and whalers set upon the Southern Ocean's marine mammals and brought some of them – including fur seals and blue whales – to the brink of extinction. Even birds were not immune from this onslaught. In the late 1800s penguins were boiled down in iron vats, so that their fat could be skimmed off and sold as an alternative to whale oil, which was becoming increasingly hard to obtain.

These activities are now largely a thing of the past, and the continent receives visitors of a very different kind. Scientists form the bulk of Antarctica's human presence, but since the late 1950s they have been joined by increasing numbers of people aboard cruise ships, who come to experience the thrill of watching ice cliffs crumble into the sea, of visiting Antarctica's islands, or of stepping ashore onto the mainland itself. Most cruise ships make their way southwards along the Antarctic

---

disaster for corals – the sediment because it clouds the water, the waste because it encourages the growth of different algae, which spread over the coral in a smothering blanket. Without careful protection the age-old growth of the coral falters and then stops altogether, and, like lights slowly blinking out, the wide variety of creatures that live and feed on the reef gradually vanish.

Because fast, long-distance travel is still relatively new, the world is only starting to see its full effects. The problems that it creates for some habitats, including coral reefs

and mountains, are already evident, and in a growing number of them efforts are being made to reduce the impact of their human visitors. But as the world shrinks the search for unspoiled wildernesses moves on, and it has now reached the very ends of the Earth.

Antarctica was the last continent to be discovered. In 1773 the English navigator Captain James Cook was the first person to cross the Antarctic Circle, but he did so in the western Pacific, where even in summer the continent is flanked by about 250 miles (400 km) of sea ice. He was convinced that

Peninsula, which extends beyond the Antarctic Circle, but some venture as far as the Ross Sea, which is the great ice-filled bay that divides the two parts of the southern continent. For those who have a taste for true adventure, summer flights across

NEW ARRIVALS *Tourists mingle with king penguins at Andrews Bay, South Georgia. Visitors could disrupt the penguins' breeding behaviour.*

the ice cap even bring the South Pole within reach, creating the ultimate excursion into a region where all human life is a recent intrusion.

With an area of about 5.4 million sq miles (14 million km²), Antarctica is about one and a half times the size of the USA, but it still sees only a few thousand visitors every year. However, Antarctic tourism is expanding rapidly, and some polar scientists are concerned about its long-term effects. The interior of the continent is a sterile

desert, with no life except for microscopic organisms stranded on the surface of the ice, but the coast has a much wider range of natural inhabitants. Lichens and algae grow on the rocks or in banks of snow, while birds and mammals use the continent's edge as a raft in the rich waters of the Southern Ocean. Despite the great length of its coastline, many of Antarctica's animals are choosy about where they breed. Adélie penguins, for example, nest in several dozen gigantic rookeries spread out

around the continent, while emperor penguins –the largest members of the penguin family – use only 30 sites, returning to them faithfully year after year.

This attachment to traditional breeding areas once made Antarctica's birds and seals easy prey for hunters, especially as they showed little concern when their enemies stepped ashore. Today the same habits make these gregarious animals the highlight of any trip to the southern continent. In some places, for example the Adélie

penguin rookeries of Torgersen Island near the tip of the Antarctic Peninsula, the arrival of tourists has become a regular event. The fat Adélie chicks, covered in a thick layer of fluffy down, look on obligingly as the cameras click and whirr, while their black-and-white parents waddle over the rocks with food for their hungry offspring.

Visitors to Antarctica are bound by a code of conduct which ensures that no one gets too close to these fascinating and photogenic animals. However, penguins have little experience of human visitors and, despite careful precautions, no one knows what impact tourism will have. On Torgersen Island researchers are currently monitoring the Adélie rookery to see if the presence of humans is causing stress to the birds and whether further protection is needed. Whatever the outcome of this work, one thing is certain: like all of the world's remaining wild places, Antarctica has become another destination for the international traveller and is destined to see a growing collection of human visitors.

BEYOND HUMAN REACH

To 19th-century whalers, who lived in cramped conditions aboard ships or in the reeking confines of whaling stations, the idea of visiting Antarctica for fun would have seemed a fantastic notion. The same can be said of the first explorers who penetrated the rain forests of the Americas, or those who ventured into the dry heart of Australia. Yet at the end of the 20th century all these places can be experienced in comfort, and most are just a few days away.

However, despite extraordinary improvements in travel, there are still places that are effectively beyond our reach. On land these include swamps, volcanoes and inaccessible caves, and also subterranean waterways and the caverns that form beneath moving ice. At sea they include the deepest parts of continental shelves, and the vastly bigger world of the oceanic abyss. Humans have managed to reach the

DISTANT DEPTHS *Light from a submersible illuminates part of the Mariana Trench in the western Pacific.*

ocean's greatest depths, but only for the most fleeting of visits. With pressures that exceed the atmosphere's by more than 1000 times, conditions here are as adverse to human life as they are on the surface of the Moon, and the slightest mechanical fault can be catastrophic.

At one time difficulties like these meant that further exploration was impossible, but with the help of modern electronics we can now investigate parts of the world that are too dangerous for us to visit in person. Our machines act as our emissaries, clambering into the vents of volcanoes or probing the deep seabed. While their operators are safely removed from the heat, cold or pressure, these self-propelled cameras peer into worlds that will probably never be seen directly by human eyes.

At present the practical value of this technology is still fairly limited. Cameras have provided information on the workings of volcanoes, and have revealed details about the final moments of ships that now lie on the seabed beneath several miles of water, but their ultimate impact may be far greater than this, because with electronic assistance we will eventually have the ability to observe every part of the biosphere – the part of the Earth's surface in which life is found. The age of exploration may be over, but what we discover during this process may be as astounding as anything that has gone before.

OUR PLANET'S CHANGING CLIMATE

The Earth's climate is characterised by change – measured in seconds and minutes, or in centuries and millennia. For most of history we have been passive witnesses, but now there is growing evidence that we are triggering changes ourselves.

Human memory is unreliable. We remember some events better than others, and we are heavily influenced by our feelings at the time. When we recall the weather of years past, we are often unable to distinguish the weather itself from the impressions it created. If we try to identify changes in weather patterns based on those impressions, the room for confusion is greater still.

Writing in the early 1770s from his home on Greenland's south-west coast, the Danish naturalist Otho Fabricius faced no such problems. Here, around the small Danish settlement of Frederikshåb – now known as Paamiut – there was ample evidence that something peculiar was happening to the climate. 'The ice spreads out more and more every year...' he wrote. 'So swift is this growth that present-day Greenlanders speak of places where their parents hunted reindeer among bare hills which are now under ice.' Fabricius did not have to rely on the memories of his neighbours to back up this statement; he had encountered very visible proof. 'I myself have seen paths running up towards the interior of the country, which were well-worn in bygone days but are now broken off by the ice's advance.'

Fabricius's remarkable observations were made towards the end of the Little Ice Age, a period of unusually cold weather that

NORTHERN GLOW *Golden leaves herald the arrival of autumn in a Norwegian valley. Farming this far north is possible due to warmth from the Gulf Stream.*

ARCTIC WARMING *In recent years, the Arctic Ocean has rapidly warmed, providing evidence for climate change.*

began in the 13th and 14th centuries, and which – after a temporary respite – reached a climax over 400 years later. For people in northern lands these were difficult times, with failed crops, wildly erratic conditions and prolonged winter freezes. During the worst periods of cold, farmers in Scandinavia fought against starvation, and in north-west Iceland fishermen worked through holes in the ice, and used horses instead of boats to bring their catches back to shore.

For many millennia people have had to cope with global changes in climate, and the Little Ice Age was simply the most recent chapter in this continuing story. These changes in climate have long puzzled climatologists, and many possible explanations have been advanced to account for them. Some theories suggest that they are triggered by purely geological events, such as bursts of volcanic activity, while others look beyond the Earth to the effects of cosmic

dust, or to the cycles in our planet's spin and orbit around the Sun. These cycles repeat themselves at intervals of between 21 000 and 96 000 years, and there is some evidence that they might be linked to changes in the Earth's average temperature.

Unfortunately, this knowledge – sketchy as it is – offers little help in charting climate change over periods of a few decades or even centuries. The great climatic cycles are like long-term trends in a stock market: they are clearly visible with the benefit of hindsight, but in the short term they are often eclipsed by many other forms of

change. Climatologists try to tease apart this tangle of trends in an attempt to identify ones that may prove to have lasting significance, but in a system as complex as the global climate this is no easy matter. Even with the aid of some of the world's most powerful computers it is hard enough to forecast local weather a week ahead, and predicting the global climate decades ahead is harder still.

As recently as the early 1970s many climatologists held the view that the Little Ice Age had paused rather than come to an end, and newspapers reported that the

TIME AND THE CHANGING CLIMATE

All over the Earth, weather and climate are constantly changing, although these changes are taking place on very different timescales. Some are rapid enough to have an immediate effect, but others take so long that we are normally unaware of their existence. At the briefest end of this spectrum are events such as thunderstorms and squalls, which often pass in less than an hour. After these come travelling areas of high or low pressure, which may linger for several days before moving on. Periodic droughts take place over a few years, as do some global events such as the Southern Oscillation, which reverses currents and weather patterns across the tropical part of the Pacific Ocean

roughly one year in four. At the next level are events that take several human lifetimes to have an effect; these include perturbations of the climate that happen within each peak of an Ice Age and within the intervals between them. The two levels of change at the longest end of the spectrum include entire Ice Ages, which last for many thousands of years, and climatic changes created by shifting continents, which take millions of years.

The human species has existed long enough to experience one complete Ice Age, but continental drift is so slow that it far exceeds our span on Earth.

PASSING STORM *A rainstorm passes over Red Rock country, Arizona. Despite their power, storms are fleeting features.*

inland is balanced by the amount of ice melting at the margins, and as a result an ice cap or glacier reaches a state of equilibrium. But if climatic conditions are in a state of change, the ice mirrors the effect. If snowfall begins to outpace melting, the ice is produced faster than it is consumed and so it begins to advance. On the other hand, if the melting overtakes the snowfall the reverse happens, and the ice starts to retreat.

This natural balancing act makes ice caps and glaciers sensitive indicators of climatic change, but interpreting their evidence is not as straightforward as it seems. This is because changing temperatures do not only affect the rate at which ice melts; frequently they also influence the humidity of the air and therefore the amount of snow that falls. In some cases warmer temperatures can increase snowfall so much that – in defiance of common sense – warming glaciers actually grow.

Faced with complications like these, most climatologists are wary of drawing

---

Earth was set for a return to colder times. However, since then the headlines have changed. Despite some voices of dissent, most climatologists now hold the view that after a century of relative stability the Earth is warming up. Unlike earlier changes in the global climate, this one may be man-made.

THE BIG THAW

In February 1995 an iceberg measuring over 1000 sq miles (2600 km²) broke free from the Larsen ice shelf and began to drift away from the Antarctic. At about the same time an ice bridge farther north melted and, for the first recorded instance since its discovery, James Ross Island became surrounded by the open sea. Two centuries after Otho Fabricius wrote about advancing

ON THE RETREAT *In north-west Greenland a retreating glacier reveals a landscape scoured by ice.*
Opposite: During the height of the last Ice Age this Chilean glacier sprawled onto low-lying ground. Today it ends high on a mountain cliff.

ice in Greenland, the world's largest ice cap seems to be on the retreat.

Ice caps and glaciers act like giant conveyor belts, carrying compacted snow downhill and – in some cases – into the surrounding sea. If the climate is stable for long enough, the amount of snow added

hasty conclusions from advancing and retreating ice. The melting of the James Ross Island ice bridge is an unusual event, as are giant cracks discovered in ice shelves close to the Antarctic mainland. When combined with the news that Antarctica's coastal ice sheets are shrinking at the rate of between 2 and 4 per cent a decade, the evidence does suggest that a major climatic change is already under way.

By comparison with continental ice caps, many alpine glaciers seem to be in an even greater hurry to withdraw to higher ground. With these, as with large ice caps, the picture is complex, and varies in different parts of the world. A few glaciers are growing, but many more are on the retreat. In the mid-1980s Chile's Balmaceda glacier gouged its way down a steep mountainside and into a deep coastal fiord. Today the same glacier stops well short of the sea, exposing a gash of brightly coloured rock scraped smooth by the ice.

This great thaw has not only affected regions in the far north and south. In a recent expedition to the island of New Guinea, Australian scientists visited Mount Jaya, a 16 500 ft (5030 m) peak where three glaciers exist just a few hundred miles from the Equator. They discovered that the glaciers have retreated by 150 ft (46 m) a year over the last two decades, and that, at current rates, one of them seems destined to disappear altogether in the near future.

ADDING UP THE EVIDENCE

Unusual weather always attracts attention, even though it may not be very significant. Even during the Little Ice Age there were brief runs of warm summers, and in the 20th century records have been broken for both summer heat and winter cold. A lack of accurate data makes it difficult to compare the present with several centuries ago, but, with the help of other kinds of evidence, climatologists can throw some light on where the climate stands today.

One way is to examine the growth rings of trees. In North America, Siberia and northern Sweden teams of scientists have analysed the rings on living larch trees, and on dead trunks that have been preserved in

waterlogged sub-Arctic peat. By measuring the thickness and density of the rings they have been able to build up a picture of growth conditions stretching back for centuries. This record is remarkably precise. It shows not only the effect of the Little Ice Age but also even the aftermath of short-lived events such as the volcanic eruption of

Mount Tambora in the East Indies in 1815. This loaded the atmosphere with so much dust that it masked some of the Sun's heat, turning 1816 into the famous 'year without a summer'.

The tree-ring record does not provide absolute temperature readings, but it does make it possible to compare one year with

WRITTEN IN WOOD *The timber from these young, fast-growing conifers will show how the climate changes in years to come. Above: Parisians enjoy a cooling shower during the hot summer of 1995.*

In many parts of the world the temperature rises and falls by at least 11°C (20°F) in the course of a single day. Compared with that, an increase of less than 1° spread across three decades hardly sounds important. However, daily changes in temperature involve relatively small amounts of heat, whereas a rise in the average temperature of the entire surface of the Earth, including the oceans, even by a fraction of a degree, involves energy on a different scale. If the present rate of increase is sustained, as many climatologists expect, the additional energy could create one of the greatest ecological changes our planet has ever seen.

If the Earth really is warming up, what has caused this rapid change in the global climate? It is unlikely to be linked with short-term changes in the amount of energy produced by the Sun; instead, it almost certainly hinges on what happens to that energy once it reaches the Earth.

Images of the Earth seen from space show a planet that is bathed in light. In fact, the Earth is visible only because it

another. The evidence gathered from hundreds of trees suggests that the 20th century has been the warmest for more than 1000 years.

Since the middle of the 19th century improved thermometers and more reliable record-keeping have produced much more detailed information about the weather, so

accurate comparisons can be made with the present day. Results published in 1995 showed that the ten hottest years since these records began occurred in either the 1980s or the early 1990s. 1995 itself turned out to be the warmest year recorded, with an average surface temperature 0.4°C (0.7°F) above the figure for the previous 30 years.

REFLECTED LIGHT *Satellite pictures of the Earth such as this are possible because the Earth reflects so much of the light received from the Sun.*

catches the Sun's light and reflects much of it back into space. Clouds and ice reflect up to 90 per cent of the light that strikes them, while the continents and oceans reflect far less. Taken as a whole, the Earth shrugs off nearly a third of the solar energy that it receives, making it twice as reflective as Mars, our nearest planetary neighbour, and over four times as reflective as the Moon.

The energy that escapes being reflected is absorbed either by the atmosphere or by the land and oceans and is then transformed in a variety of ways, some of it driving the movement of the air and oceans, and some of it making water evaporate. A large amount of it is radiated from the ground, only to be absorbed and re-radiated by the atmosphere above. But whatever happens to energy during its time on Earth,

and however long it shuttles between ground, air and water, its ultimate fate is always the same: it drains away into space.

The Earth's energy account, like a stable glacier's ice account, is normally in a balanced state. The amount of energy that it receives is the same as the amount that it loses to space, just as the amount of ice that builds up in a stable glacier is the same as the amount that melts away. But within this overall state of balance, subtle changes occur. If some factor on the Earth's surface alters, its energy flows subtly shift. This shift continues until a slightly different balance is struck and a steady state returns.

Human activities have changed energy flows on Earth in many ways, but hardly any of them have the potential to exert truly global effects. One of the few that might do this is our role in creating dust and other fine particles, which are carried high into the air. By making the atmosphere more opaque, these particles help to scatter more solar energy back into space, so slightly less manages to reach the ground. Volcanoes release dust in massive amounts, but even

after allowing for their eruptions the invisible veil of man-made particles has increased rapidly in recent years, imperceptibly weakening the heat of the Sun.

If dust alone were at work, the Earth's temperature would be falling. As the reverse is true, something else must be working in the opposite direction. After years of speculation, the identity of that something now seems beyond reasonable doubt.

THE CHANGING AIR

The Irish physicist John Tyndall is remembered mainly for a discovery that can be demonstrated with a glass of water and a few drops of milk. If the milk is stirred into the water, and a flashlight shone through the mixture in a darkened room, floating particles scatter the light sideways and reveal its otherwise invisible beam. But in a scientific paper published in 1863 Tyndall announced a different discovery that has a much wider relevance for us today. He had found that carbon dioxide and water vapour are transparent to light, but that they are not transparent to heat.

Tyndall's breakthrough helps to explain why the Earth – unlike the Moon – stays

LIVING THERMOMETERS

Many plants and animals are very sensitive to changes in temperature, and their distribution is affected by variations in climate. In North America one species of butterfly – Edith's checkerspot (*Euphydryas editha*) – used to be commonly found from Canada as far south as Mexico. The southernmost checkerspots are now dying out because the climate has become too warm and dry for their normal foodplants to survive. By contrast, the checkerspots that live farther north or at a high altitude continue to thrive.

reasonably warm after the Sun has set. During the day sunlight shines through the atmosphere and warms the ground, and after dark the ground re-radiates this heat back into the sky. But because the Earth is cooler than the Sun, it emits its heat in longer wavelengths, known as radiant heat or infrared radiation, which we can feel but cannot see. This energy is often absorbed by clouds, but even on a cloudless night it does not have a free run out into space. Carbon dioxide, water vapour and other gases block its path, forming an invisible jacket that keeps the Earth's surface about 30°C (55°F) warmer than it would otherwise be.

THE GREENHOUSE EFFECT
When the level of greenhouse gases in the atmosphere increases, more of the Sun's infrared (heat) rays are blocked by the atmosphere. As a result, the surface of the Earth warms up.

Known as the greenhouse effect, this blocking of infrared radiation is essential to much of life on Earth. The atmosphere does not block all infrared wavelengths – if it did the Earth would be far warmer than it is – but it does make it more difficult for radiant heat to escape. Because life depends on chemical reactions, and because chemical reactions are accelerated by heat, the greenhouse effect keeps the Earth's ecosystems running at their often brisk rate.

Tyndall did not pursue his discovery to its global conclusions, but in 1896 Svante Arrhenius, a Swedish physicist and chemist, did. Arrhenius was interested in the idea that reduced carbon dioxide levels might have caused the ice ages, but he also explored the implications of a change

in the opposite direction. He lived in an age of burgeoning industrial production, when the use of fossil fuels was on the increase. As he pondered the puzzle of climate change he realised that the carbon dioxide released by burning fuels might eventually alter the atmosphere's composition and increase its ability to hold on to heat. As things have shown, his supposition was correct.

In the natural world carbon dioxide is a remarkably scarce gas. For every one unit of carbon dioxide in the air there are about 700 units of oxygen and over 2500 of nitrogen; both are so abundant that it would be difficult for us to alter the level of either of them. Carbon dioxide, however, is another matter. Because it is one of the atmosphere's trace ingredients, its level is

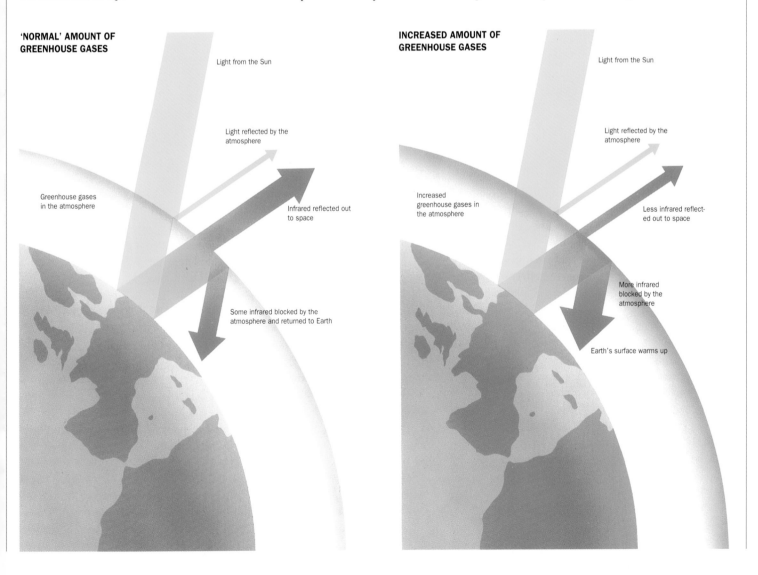

'NORMAL' AMOUNT OF GREENHOUSE GASES

Light from the Sun

Light reflected by the atmosphere

Greenhouse gases in the atmosphere

Infrared reflected out to space

Some infrared blocked by the atmosphere and returned to Earth

INCREASED AMOUNT OF GREENHOUSE GASES

Light from the Sun

Light reflected by the atmosphere

Increased greenhouse gases in the atmosphere

Less infrared reflected out to space

More infrared blocked by the atmosphere

Earth's surface warms up

within the reach of man-made change.

By drilling into polar ice, climatologists can retrieve bubbles of ancient air that record the make-up of the atmosphere long ago. Bubbles trapped in ice between AD 1000 and 1750 reveal that carbon dioxide levels were fairly steady during this period, at about 280 parts of the gas for every million parts of air. But after 1750, and the beginning of the Industrial Revolution, the bubbles tell a different story. By 1850 the carbon dioxide level stood at nearly 300, and by 1900 it was 310. Just as Arrhenius predicted, the carbon dioxide level had started to take off.

In 1957 the world's first carbon dioxide monitoring station was set up on Mauna Loa in Hawaii, and a continuous sequence of readings came on stream. The results for the late 1950s show that the carbon dioxide level had already reached about 315 parts per million, and it has continued to climb ever since. At the end of the 20th century the level stands at about 350 parts per million – an increase of 11 per cent in less than 50 years.

If the Earth's temperature had risen in step with this change, the greenhouse effect

GREENHOUSE GASES

Efforts to curb global warming focus mainly on reducing the level of one 'greenhouse gas' – carbon dioxide. Although this is the chief agent of global warming, several other gases play an important part in helping the Earth to retain heat. One of these is nitrous oxide, which is produced by bacteria in soil and also by burning fuels. Another is methane, which is formed by bacteria in swamps and marshy ground, and also by grazing mammals. A third group of these gases is made up by the CFCs, which until recently were widely used as refrigerants and propellants in aerosols. Unlike nitrous oxide and methane, CFCs do not exist in nature. Compared with carbon dioxide, the warming effect of these extra greenhouse gases can be extremely high. Methane, for example, is

CYPRESS SWAMP *Swamps are a major source of methane. It forms when plant remains break down under water, where oxygen cannot reach them.*

approximately 25 times more effective in trapping heat than carbon dioxide, while some CFCs are 20 000 times more effective, so their impact far outweighs their relative scarcity in the atmosphere.

CFCs are currently being withdrawn from use, but methane is a growing problem in the drive to stabilise the Earth's temperature. Its level in the atmosphere has more than doubled since the 18th century and is

currently growing at about 1 per c year. Most of this increase has be produced by the expansion of rice paddy fields and livestock rearing from bacteria living in rubbish tips Some also comes from decaying plants that have been drowned by hydroelectric reservoirs. Remains such as tree trunks can take seve centuries to rot away, and through this time they release methane in the air above.

might have become a major news item at the turn of the last century instead of at the end of this one. However, the Earth's climate, as always, behaves in complex and often confusing ways. Between 1880 and 1940 global temperatures rose by about 0.25°C (0.45°F), and some climatologists did speculate about an increase in the greenhouse effect, and its possible role in the American dustbowl disaster of the 1930s. But between 1940 and 1970 the pattern was reversed, and the world cooled by about 0.2°C (0.35°F). Most climate experts followed suit, and instead of studying global warming they

focused on global cooling instead. S the 1970s temperatures have been ri once more, with carbon dioxide le surging ahead. While there is no direct between global temperature and car dioxide levels, this time the coincidene difficult to dismiss.

THE UPS AND DOWNS OF GLOBA WARMING

To Otho Fabricius and his neighbou 18th-century Greenland, global warm might have sounded like good news. T survived in a land where summer is on few weeks long, and endured eight-mo winters without the benefit of modern I ing. In such conditions, any extra war would have been an attractive prospect.

In parts of the world today, global w ing is sometimes greeted with just this of enthusiasm. Scotland, for example, not have the punishing climate of wes Greenland, but it is often chilled by w

THE FROZEN PAST *Ancient blocks of ice contain snow, air and atmospheric dust that provide a detailed picture of the climate in the distant past.*

blowing down from the far north, and summer temperatures average only about 17°C (63°F). An increase of just 1°C (2°F), which could well come about in the next three decades, would certainly help to make life there more comfortable. In towns like Ushuaia, near the stormy southern tip of South America, global warming would probably be even more welcome. Although it is at the equivalent latitude to Scotland its climate is cooler still, and even in midsummer the thermometer rarely manages to struggle above 10°C (50°F).

Even the most ardent environmentalists agree that global warming may bring some advantages to places like these, and that rising temperatures are preferable to ones heading in the other direction. At high latitudes the extra warmth would lengthen the growing season by increasing the number of days free of frost, and throughout the world as a whole the potential rate of plant growth would be increased, because plants use carbon dioxide as one of their main raw materials. However, global warming threatens some harmful changes in its wake, and a major international effort is being made to bring carbon dioxide emissions under control.

One of the most important of these changes is the prospect of rising sea levels, a phenomenon that is already under way. During the 20th century, the sea level has risen by 6 in (15 cm), and a further rise looks inevitable, whatever steps are taken to reduce carbon dioxide levels. Sea levels have risen and fallen by dozens of feet throughout history, so this kind of change is nothing new. However, previous fluctuations happened before our species developed an urban lifestyle. It would have been easy for nomadic hunters to abandon their traditional campsites and move to higher ground; abandoning modern cities on coasts and estuaries is a different proposition.

The rise in sea levels is often portrayed as the result of melting ice, but recent

THREAT FROM THE SEA *For low-lying islands such as the Maldives, any rise in sea level spells potential disaster.*

research suggests that this might be a relatively minor factor, and the expansion of the seawater itself is likely to be more important. Water is a very unusual liquid, because at low temperatures it shrinks as it gets warmer. Once it has reached about 4°C (39°F) it starts to behave 'normally' and expands as it heats up. The increase in volume is quite small – less than 0.001 per cent for a warming of just 0.5°C (1°F) – but, given the vast dimensions of the oceans, its effect at the surface is pronounced. Compared with most liquids, water is also unusually good at storing heat. This means that once the oceans have warmed up they stay warm for a long time, so any rise in sea level is difficult to reverse.

Over the last two decades the debate about how far the sea will rise has produced a stream of figures, some of which contrast sharply. Early estimates talked of a rise of 11^1/$_2$ ft (3.5 m) by the year 2100, which

would have wiped half of Florida and the Netherlands off the map, and flooded most of Bangladesh. The calculations today are much more conservative – an official US estimate in 1995 produced a 100 year figure of just 13^1/$_2$ in (34 cm) – but even this may have important consequences for many shoreline cities. For low-lying coral islands such as the Maldives, which are often less than 3 ft (1 m) above the sea, the implications are alarming.

For those who live on higher ground, the problem of rising sea levels is of less immediate interest. But the extra heat created by global warming will alter the Earth's weather systems and, because these reach into the heart of every continent, climate change will be a reality that none of us can escape. In this meteorological lottery some areas will benefit from increased warmth and rainfall. The Middle East, for example, may well become wetter and more fertile, as

FAILING RAIN *These hillsides in southern Spain show the effects of a five-year drought. Further droughts are likely as dry air moves northwards from Africa.*

may parts of Australia and the southern fringes of the Sahara. But for every winner there is likely to be a loser. Southern Europe, which has already suffered repeated droughts in recent years, will see those droughts becoming more severe as dry air shifts north. Central Russia is likely to become more arid, as is the grain-producing region of North America.

CLIMATIC FLIPS

As knowledge about the global climate has advanced, it has become clear that the Earth has many complex mechanisms that govern the process of change. Increased warmth, for example, boosts the amount of cloud, which in turn reduces the level of incoming solar heat. In a different mechanism the seas absorb carbon dioxide from the air, and the more carbon dioxide there is in the atmosphere, the faster this absorption occurs. This helps to put a brake on the greenhouse effect by reducing the speed of change.

These two examples are both negative feedback systems, so-called because they tend to reverse any changes that occur. However, the Earth also has feedback systems which work in the opposite direction, exaggerating any change that occurs. When the Earth warms, for example, more water evaporates into the air. This water vapour helps to trap more heat, making the Earth warmer still. Over the past century negative feedback systems seem to have had the upper hand, so the changes triggered off by mankind have been less marked than might have been expected. But this fortunate state of affairs may not continue. What would happen if we inadvertently bolstered the Earth's positive feedback systems, so that global warming was speeded up? How far might the climate change before it once more reached a balance?

In 1993 some startling new evidence came to light which shows just how quickly the Earth's climate can make the transition from one state to another. Working in one of the world's most remote places, in the centre of Greenland, an international team of scientists spent two years boring a hole through the ice cap to the bedrock nearly 2 miles (3.2 km) below. The ice cores (cylinders of ice) retrieved from this hole

MOUNTAINS UNDER ICE
Greenland's ice cap has existed for over a million years. Each year, fresh snow preserves a record of changing conditions.

ARCHIVES IN ICE

When snow forms, it contains two types, or isotopes, of oxygen – oxygen 16 and oxygen 18. Oxygen 18 is the rarer of the two, but the exact ratio between the isotopes depends on the temperature of the surrounding air. The colder the air, the less oxygen 18 is built into each snowflake. By extracting ice cores from glaciers and ice caps, meteorologists can examine snow that fell long ago. The ratio between its oxygen isotopes gives an accurate record of temperature changes.

form a continuous record that dates back over a quarter of a million years. When analysed they reveal not only how much snow fell in each year but also what the average air temperature was when each flake of snow was formed. These were the first cores to cover two complete ice ages and three interglacials, or warm periods in between.

The cores' bands of ancient snow make it clear that, far from being under siege by the weather, we are living in clement times. During the present interglacial, which has lasted about 10 000 years, the global climate

has been unusually stable, despite the Little Ice Age and other departures from the norm. Climatologists once assumed that the previous interglacial – which was slightly warmer than our own – followed the same pattern. However, the Greenland ice cores show that this was probably not the case. Instead of being relatively static, temperatures sometimes plunged by as much as 10°C (18°F) in two or three decades before inexplicably recovering, and when the interglacial finally came to an end, about 114 000 years ago, they seem to have fallen by about 14°C (25°F) in just ten years.

This kind of instability has been aptly compared to a ball bearing rolling about in a saucer. Given a moderate push, the ball bearing climbs close to the edge of the saucer but eventually drops back towards the centre. Given a harder push, it suddenly falls over the edge. It is an alarming possibility, and as yet nobody knows how close we are to the edge.

CIRCULATING HEAT

On the wave-pummelled coasts of northwest Europe, stranded voyagers from the tropics can sometimes be spotted where they have been washed up along the

shore. These voyagers are large seeds, about 2 in (5 cm) across, which are produced by plants that grow in the faraway Caribbean. Known as sea beans (*Entada gigas*), they have no hope of germinating as far north as this, but their extraordinary odyssey demonstrates the power of the Sun and wind to keep seawater constantly on the move.

Currents operate in all the Earth's seas and oceans, and help to spread the heat that is generated by tropical sunshine. Warm surface currents carry water away from the Equator, and cooler and much deeper currents creep along the ocean floors, replenishing the water that has been displaced. Radiocarbon dating shows that throughout the world's oceans as a whole, seawater flows through a giant circulatory system that takes about 1000 years to complete.

It is easy to imagine that this circulatory system has existed as long as the oceans themselves, and that it is fixed by the position of the continents and the shape of the sea floor. However, according to some recent discoveries, this may not be true. Far from being constant, it is possible that some currents could abruptly change, producing exactly the kind of positive feedback that might propel the Earth's climate into a new and quite different state.

Among the key features of this global circulation are areas where the sea's surface regularly freezes and thaws each year. Sea ice contains very little salt, so when it forms the water beneath it becomes saltier than usual. This cold and heavy saline water sinks to the seabed, forming a pump that pushes water towards the tropics, where warm currents are born. If something alters the effectiveness of such a pump, the consequences could be far-reaching.

In 1994 a leading oceanographer reported that one of these pumps – a tongue of ice off eastern Greenland – no longer seemed to be operating with the strength it once had. During a succession of warm winters much less sea ice has formed, and so less saline water has sunk to the seabed. In 1995 a group of scientists working in the eastern Mediterranean found deep-water changes of a different kind. Here the directions of the currents have changed, and the seabed water is now saltier than before. Oceanography is a relatively young science, and it may be that these changes are simply natural oscillations that have not been observed before. But if they are linked instead to a rise in the Earth's average temperature, they would certainly have the power to turn a minor climatic change into one that is far more significant.

Few parts of the Earth would feel that change more than north-west Europe. Thanks to the Gulf Stream, which propels sea beans across from the Caribbean, these shores are bathed by warmth that is created far away. By the time the seawater reaches Europe its surface temperatures are far cooler than in the tropics, but it still contains sufficient heat to transform the climate.

As the water nears the end of its journey northwards it rounds Norway's Arctic coast and fades away in the Barents Sea close to the Russian city of Murmansk. The Gulf Stream's cargo of heat ensures that Murmansk remains ice free all year, while most other ports as far north as this are locked solid in winter by a crystalline sea.

If the Gulf Stream weakened, and failed to reach northern Europe at all, the climate would soon become like that of north-east Canada. Rivers and lakes would freeze, tender plants would die, and most types of farming would be impossible. White Christmases – a rare treat in southern Britain – would be as reliable as the setting Sun.

PLANNING THE FUTURE

At present, climatic flips and failing currents are the stuff of scientific speculation rather than looming hazards that we will soon have to confront. But even if global warming does not produce such dramatic developments, it undoubtedly promises to

DISTANT WARMTH *The Orkney Islands are on the same latitude as the southern tip of Greenland, but the Gulf Stream gives them a milder climate.*

make life more difficult. Insurance companies are already trying to predict the cost of an increasingly volatile global climate, and with more heat being stored up at the Earth's surface that uncertainty can only increase.

The answer to these problems is to reduce the atmosphere's level of greenhouse gases until we approach something close to their level in the 1860s, when John Tyndall first realised the potential that they have. The Intergovernmental Panel on Climate Change (IPCC), which was set up in 1990, has agreed on some steps in this direction, but making the reduction will be neither a fast nor an easy process.

We have added about 300 billion tons of carbon dioxide to the air since the Industrial Revolution began, and today this gas is an inescapable by-product of many of our sources of energy. Carbon dioxide emissions can certainly be reduced by using energy more wisely, and by generating it from renewable sources. It can also be removed from the atmosphere directly by planting trees, which use it to build their wood.

However, during the 20th century we have also liberated other greenhouse gases that have a much longer potential life span in the atmosphere. Few of these can be withdrawn from the air by plants, and

ON THE MOVE *Water surges down the Nahanni River in north-west Canada. Wherever water moves, heat energy also moves, shaping the climate.*

most will have to be broken down or removed through the atmosphere's natural chemistry.

There is little doubt that their presence in the air above us will trigger global warming for many decades, until curbs on greenhouse gases start to have an effect. During that time the Earth will see many changes, but no one can yet say exactly how they will affect its population.

NATURE IN RETREAT

In nature, life is a constant challenge. Some species thrive; others fail and slowly disappear. In recent times, however, the challenge has changed. As well as confronting the hazards of daily life, creatures have to face new competitors: ourselves.

In September 1914 newspapers in Cincinnati, Ohio, carried an unusual obituary. The deceased was female and had died, aged 29, after spending most of her life alone. However, far from suffering from neglect, she had been well looked after and had generally been in the best of health. Her name was Martha, and she was the world's last passenger pigeon – a species that was once the most numerous bird on Earth. Martha's death in captivity closed one of the most sombre chapters in the history of North American wildlife. During the early 19th century passenger pigeons were extraordinarily abundant, and they migrated across the continent in vast flocks that could take several days to pass by. Photography was then in its infancy, and no one gave much thought to taking pictures of wildlife, but eyewitness accounts from the time give some idea of the huge numbers of birds that could be seen. The artist John James Audubon estimated that one flock contained over a billion

PORTRAIT FROM THE PAST
*Today the quelea (opposite)
is the world's most abundant
bird. Two centuries ago the
now extinct passenger pigeon –
depicted (right) by the artist
John James Audubon – was
more numerous still.*

birds, while the ornithologist Alexander Wilson saw another that contained at least twice that number. This gigantic assemblage of pigeons equalled the entire population of the most abundant species of bird alive today – an African finch called the red-billed quelea (*Quelea quelea*).

However, these enormous migrations were not to last. For hunters, a bird that ate, moved and bred in gigantic groups was an irresistible target, and many millions were blasted out of the skies or out of their treetop roosts. Passenger pigeons were hunted remorselessly throughout the 19th century, and by the 1870s the species was already in serious difficulties. In 1871 the pigeons put on one of their final shows of strength, with over 130 million birds nesting in one part of the state of Wisconsin. However, while impressive by ordinary standards, this was just a faint echo of the mass nestings of the past. Within 50 years the unthinkable had happened: an animal that had once been almost overwhelmingly abundant had become extinct.

Despite the obvious damage inflicted by hunting, the reasons for the passenger pigeon's sudden demise are still far from clear. How did a bird that still numbered many millions in the 1870s disappear in the space of a few decades? Why did the pigeon fail to recover when hunting was brought under control? Was its original population steady, or was it already in decline before hunting and deforestation began? As biologists investigate our effects on the natural world, the answers to questions like these may play an important part in preventing other species from sharing the passenger pigeon's fate.

EXTINCTION AND SURVIVAL

Nature is remarkably resilient. Given the chance, plants and animals are quick to reclaim lost ground and are always ready to exploit any new opportunity that presents itself. Many forms of life – from dandelions to butterflies – have the potential to produce large numbers of offspring, and if conditions are in their favour they can multiply a hundred or a thousandfold in the space of a single year.

Given this resilience, it is perhaps not surprising that the world's natural inhabitants have sometimes been treated as if they were inexhaustible. In the early 19th century hunters took it for granted that the supply of passenger pigeons was unlimited, just as other hunters in later years imagined that there was a limitless supply of grazing mammals on the African plains. This assumption was shared not only by people who pursued animals for food but also by many who had a professional interest in the natural world. Nineteenth-century entomologists trapped

PLANT PLUNDER *After years of over-collection in the 19th century, the exquisite Himalayan orchid* Vanda caerulea *is probably extinct in the wild.*

per plant is not unusual – but each of these dust-like specks faces an extraordinary struggle before it can become an established plant and produce seed itself. In the case of a tree-dwelling species such as *Vanda caerulea* the seed must land on a suitable branch where it is exposed to just the right amount of daylight, and once it germinates it must secure itself to its support using special aerial roots. Each plantlet then has to collect nutrients from the dust and debris that settle on the branch, and gather water from the moisture that trickles down the tree. Altogether it takes several years for the plantlet to become established. By both removing adult plants and cutting down trees, Victorian plant collectors disrupted the all-important supply of seed, and the subsequent unsustainable harvest brought the species to the very edge of extinction.

HUNTING AND FISHING

In the case of the passenger pigeon a more complex mix of forces was probably at work. Each female pigeon usually raised just one nestling at a time but could breed several times a year. The species as a whole seemed to depend on its immense numbers for survival, and often hundreds of nests would be crammed side by side, with their occupants just beyond pecking distance of their neighbours.

Exactly why the pigeon needed to be so gregarious is not known, but huge flocks may have helped the birds to swamp their many natural enemies or to make the best use of their food supply. Once their numbers had been whittled down by hunting, and by the destruction of the trees that provided food and somewhere to nest, the population seems to have dropped below the critical figure needed for survival. Even though the guns fell silent the remaining birds failed to breed normally, and from that moment the species was doomed.

vast numbers of butterflies so they could be catalogued and filed away in museum cabinets, and if a species showed the slightest variation in colour or patterning they diligently collected examples of each type. Ornithologists shot birds or trapped them in nets, while plant collectors frequently stripped forests and hillsides of anything that could be shipped home, cultivated and sold to enthusiasts.

One collector, the British botanist Sir Joseph Hooker, combed the foothills of the Himalayas for *Vanda caerulea*, a species of orchid that was reputed to be the most beautiful in the Indian subcontinent. In his

notebooks Hooker mentions that in one particular area his assistants collected so many of the plants, often by felling the trees on which they grew, that seven wellladen men were needed to carry them all. Far from keeping their location secret, Hooker inadvertently helped other planthunters to home in on the best sites, and these collectors were so thorough that even today hardly any plants remain.

The story of this rare orchid highlights the dangers that can confront any species that is harvested or hunted by humans. Compared with most plants, orchids produce vast quantities of seed – over a million

The extinction of the passenger pigeon emphasises a fact that has implications for many forms of wildlife today: a species can become fatally endangered long before its last living representatives meet their end.

Throughout the history of life on Earth new species have gradually evolved, while others have slowly lost the struggle for survival and have disappeared. At various points in the past – most recently about 65 million years ago when, scientists believe, an asteroid may have smashed into the Earth – natural cataclysms have swept away a large proportion of species in a relatively short period of time. The process of evolution has invariably made good these losses, but in each case millions of years have passed before the species total has risen to the level at which it stood before.

Compared with these immensely distant events, changes brought about by human beings are much more recent. Some palaeontologists believe that human hunters eradicated some species of large land animals in Africa and Asia over 100 000 years ago, and it is possible that humans played an important part in the disappearance of large North American mammals such as mammoths between 10 000 and 12 000 years ago, although this has long been a subject of scientific debate. What is certain is that in more recent times humans have had a key role in determining the fortunes of many plant and animal species. Since the year 1700 over 100 species and subspecies of mammal have become extinct, along with perhaps 200 types of birds. Some of these animals would have died out of their own accord, but many

would still exist today if human intervention had not triggered their demise.

In the early days of human influence on the natural world, hunting was by far the most important threat to the existence of wild animals. In the more recent past it has accounted for mammals such as Steller's sea cow (*Hydrodamalis gigas*), a relative of today's manatees and dugongs, the last of which met its end on Bering Island in 1768; the African bluebuck (*Hippotragus leucophaeus*), which became extinct in 1800; and the quagga (*Equus quagga*), an African

REDUCED RANKS *Manatees survive in the Americas and West Africa, but their Arctic relative, Steller's sea cow, has been wiped out.*

THE QUAGGA *Seemingly part-horse and part-zebra, the quagga lived in southern Africa. This drawing shows one of the last survivors, which died in a European zoo.*

relative of horses and zebras which died out in 1883. No species of whale has become extinct in modern times, but several – including the blue whale (*Balaenoptera musculus*) and the northern and southern right whales (*Eubalaena*) – have been hunted to the verge of extinction, and of these three the northern right whale may now be beyond the point of recovery.

Birds that have died out solely through hunting are relatively few. One of the foremost was a flightless North Atlantic sea bird called the great auk (*Pinguinus impennis*), which was the Northern Hemisphere's equivalent of the penguins. It bred on scattered low-lying rocky islands off Newfoundland, Iceland and Scotland, where it was beyond the reach of foxes and other ground-dwelling predators. Unfortunately these retreats were not safe from passing sailors, and during the breeding season auks were gathered up in their thousands. The last two were killed in 1844, ironically not for their meat but for their skins, which were sold to a collector. Later in the 19th century many other birds were hunted for their feathers. Many species of hummingbird from South America succumbed to this onslaught and they are known to science only from stuffed skins that were exported to adorn hats.

Today the only forms of animal life that are still hunted on a large scale are fish. For centuries the human impact on sea fish was slight, because boats were small and fishing equipment primitive. Fishermen learned how to exploit the regular migrations of species like tuna, but because the oceans are so vast they harvested only a tiny proportion of the fish that were available. Improvements in fishing technology since the 1950s, including sonar and satellite positioning systems, have made it increasingly difficult for fish to escape the net. In a few decades the world's fish catch has more than quadrupled, and there are now clear signs that some fish are being caught more rapidly than they can replace themselves. Large fish are increasingly difficult to find, and several important fisheries, such as the cod fishery of the Grand Banks off eastern North America and the herring one of the North Sea, have suddenly collapsed – a situation that is disturbingly reminiscent of the last years of the passenger pigeon.

Marine biologists are still learning about the complex ecology of sea fish and the factors that control their numbers. Studies of fish scales preserved in the seabed mud off California show that some species undergo natural cycles of abundance and depletion even before they are fished. These cycles may explain why some species are harder to find than they were, but they are unlikely to be

A TALE OF TWO WHALES

The recent history of two species of whale shows how human exploitation can have a very different impact on closely related animals. The northern right whale (*Eubalaena glacialis*), which lives in the North Atlantic and North Pacific, was once widely hunted because it was relatively tame, lived close to the shore and floated once it was dead. It yielded valuable oil and large amounts of baleen – the fibrous strips that some whales use to filter plankton while feeding – which found its way into different products from corsets to umbrellas. Originally there may have been over 100 000 right whales in the Northern Hemisphere, but by the time the species was given full protection in 1937 the

TAKING A BREATH *A rare northern right whale floats on the surface, revealing its double blowhole. Whalers once identified this animal by its twin vapour-filled spouts.*

total had dropped to a few hundred. Despite this eleventh-hour attempt to save it, the northern right whale seems to have approached its biological breaking point, and over five decades later it is still hovering on the brink of extinction. By contrast, the smaller grey whale (*Eschrichtius robustus*) has fared far better since it was protected in 1946. Grey whales became extinct in the North Atlantic in the 18th century, and by the end of the 19th century those in the eastern Pacific were also fighting for survival. However, protection produced a dramatic reversal in fortune for this species. Grey whales are now common off the west coast of North America, and the future of the species is more secure than at any time in the last 150 years.

PENGUIN OF THE NORTH
The great auk met its end on
June 2, 1844, when three
fishermen landed on Eldey,
an island off south-west
Iceland. Being flightless, the
last pair could not escape.

the only reason. For species such as cod and herring there is little doubt that today's hunters of the oceans have helped to shrink stocks to a fraction of their former size.

WINNERS IN A CHANGING WORLD

While some land animals have been eliminated by hunting, others have been brought to the point of extinction by human intervention of a different kind. These include the introduction of predators and competitors, such as cats, goats and pigs, and also the peculiarly human habit of collecting wild animals as pets. More recently, pollution has been added to the list of hazards that wildlife has to face.

However, since the dawn of agriculture one activity has developed an overriding importance in the way we influence the world of nature. This is habitat change, and its effects can now be seen in almost every landscape on Earth. As the human world has expanded globally, the natural one has been rolled back. Forests have been cut down, wetlands drained and grasslands ploughed up. In some places entire natural habitats have completely vanished, while in many others they have shrunk to a fraction of their former size. The impact on wild plants and animals has been profound.

When humans alter a habitat there are winners as well as losers; for example, the skylark (*Alauda arvensis*), whose airborne song has inspired many British poets and composers, would once have been a scarce bird in the British Isles because it needs open grassland in which to nest and feed. Until the advent of agriculture most of the British Isles were covered with forest, so the skylark would have found few places in which to breed.

Studies in Finland and California have shown that grassland birds are not the only ones that have benefited in this way. When natural habitats are replaced by mixed farmland and suburban gardens, the total weight – or biomass – of birds can actually increase. A similar story can be told of some

plants. The jewel-like bee orchids (*Ophrys*) of southern Europe, which lure male pollinating insects by imitating the shape and smell of their female mates, thrive in sunny, open ground. During the days when the Mediterranean region was largely covered by trees, these fascinating plants would have been relatively rare and would have been confined to crumbling slopes where trees could not grow. After centuries of deforestation, large parts of southern Europe

ROADSIDE FLOWERS *Roads destroy some habitats, but verges (above) create new ones for many wildflowers. European bee orchids (right) thrive in open, sunny landscapes, most of which have been created by deforestation and browsing livestock.*

provide them with an ideal habitat. Many other plants also stand to gain from human changes to natural habitats. Grazing animals keep woody plants at bay, so natural grassland wildflowers can survive, and roadsides provide just the right conditions for plants that need bright sunshine and good drainage. They also create an ideal habitat for many small mammals, which inadvertently provide food for scavenging animals when they wander into the path of oncoming cars. Birds such as the American turkey vulture (*Cathartes aura*), the European carrion crow (*Corvus corone*) and magpie (*Pica pica*) regularly feed on this unexpected harvest.

The beneficiaries of changing habitats also include pigeons, starlings, and mammals such as brush-tailed possums, raccoons and foxes, all of which have managed to adapt to urban life. However, these animals are a relatively privileged few. For the majority, habitat change brings difficulties that may be very hard to overcome.

THE DANGERS OF SPECIALISATION

Humans are remarkably adaptable, and are unmatched in their ability to thrive in a wide range of different habitats. A handful of wild animals, such as the grey wolf (*Canis lupus*) and the cougar or mountain lion (*Felis concolor*), can survive in more than one kind of habitat, but most species depend on a particular kind of environment and on the other forms of life found within it. In many cases the links between living things and their habitats are extremely strong, making it possible to predict what kinds of life are likely to be found in a particular place.

Many of these links are the result of specific chemical needs. For example, crayfish – freshwater relatives of lobsters – need large amounts of calcium carbonate to build their hard body cases. They are most numerous in streams that flow over calcium-rich rocks, such as limestone and chalk, but are rarely found in water that flows over rocks such as granite, which is calcium-poor. By contrast, plants such as heathers (*Erica*) and rhododendrons have exactly opposite chemical needs. They thrive in acidic soil and are 'calcifuge', meaning that they cannot tolerate too much calcium. This explains why they thrive in some gardens but turn yellow and die in others.

For some animals a habitat offers a physical resource that is found nowhere else. For ant lions, which are the grubs of insects that look like small dragonflies, that resource is dry, loose soil. Each grub excavates a steep-sided pit 1 in (2.5 cm) or so deep and hides at the bottom of it with its jaws just visible. Ant lions feed by using the insect equivalent of artillery. Their 'shells' are grains of soil, and they flick these at passing ants, making their quarry tumble into the pits where they are seized and eaten. This feeding technique works superbly in dry open places, where the sides of the pits crumble away when an ant tries to escape, but is useless in regions where it frequently rains. Here the ground often becomes sticky and overgrown, and in these conditions ant lions would soon starve to death.

Luckily for ant lions there is no shortage of this kind of habitat, and in warm parts of the world the ground is often pockmarked with their unusual but highly effective traps. However, for many other animals reliance on a particular habitat has turned out to be a much less successful evolutionary gamble. One species that has suffered as a result is the ivory-billed woodpecker (*Campephilus principalis*) of the eastern United States, a striking black and white bird which most ornithologists believe is now extinct. With a body length of almost 20 in (50 cm) this is – or more probably was – the largest woodpecker in North America. Because of its exceptional size it needed several square miles of forest to provide it with food and large dead trees in which it could excavate its nests. Until the arrival of European colonists, trees like this were relatively common, but when forests were cleared the woodpeckers soon disappeared.

New trees eventually replaced much of the original forest cover, and other woodpeckers managed to move back and set up home. However, this regrowth did not suit the ivory-billed woodpecker, and the species has not been reliably reported in North America since the 1940s.

In any habitat, animals are bound together by complicated sets of relationships among predators, partners and prey, and even the subtlest changes can affect many different species. One remarkable example of this concerns a European butterfly called the large blue (*Maculinea arion*), which until 1979 bred in small areas of grassland in southern England. The large blue belongs to a family of insects that as caterpillars show some intriguing relationships with ants. Most blue caterpillars produce droplets of a sugary fluid which ants use as food, and in return the caterpillars gain protection provided by their aggressive guardians. However,

the caterpillars of the large blue have a more sinister side to their nature. They start life feeding on wild thyme (*Thymus praecox*), but at an early stage in their lives they trick ants into 'adopting' them and carrying them into their nests. Once safely underground they turn into predators and feed voraciously on the helpless ant grubs.

In southern England the wild thyme and ants on which the large blue depends are found only in grassland that has never been treated with fertilisers or herbicides, and is closely cropped by grazing farm animals or by rabbits. Before the First World War this kind of grassland was quite common, but changes in farming practices have since made it much more scarce. To make matters worse for the butterfly, the myxomatosis virus was deliberately introduced into England from South America in the 1950s in an attempt to control spiralling rabbit numbers. It had an immediate and

THE FOREST THAT CAME BACK TO LIFE

When biologists try to assess our impact on the natural world, they often face difficulties in distinguishing man-made changes from ones that occur in nature. One of the most important demonstrations of this problem occurred in the early 1980s, when large areas of forest in Germany appeared to be affected by a mysterious condition called *Waldsterben* or forest death. The leaves of evergreen conifers turned yellow and fell, and ecologists concluded that pollution was slowly poisoning the soil on which they grew. There was widespread alarm, and newspaper headlines announced that by 1990 much of the country's forest cover would be dead. However, by the mid-1990s it became clear that these dire predictions were wide of the mark. Although some trees are unhealthy, the rate of tree growth in Germany's forests has actually

increased, and there are no signs of the catastrophe that was widely anticipated. Many forest scientists now acknowledge that their projections turned out to be wrong. There is no doubt that Germany's trees have been affected by pollution, for example through acid rain, but it seems that *Waldsterben* was more the result of climatic changes, which brought warm and dry weather from the mid-1970s onwards. The conifers reacted to this change by prematurely shedding their leaves, giving the impression that much of the forest was doomed. Today large areas of Germany's forests show a record volume of wood and have outgrown past problems. The confusion over

Waldsterben and the factors that caused it shows just how difficult it can be to discover what effect we have on the world around us.

BLACK FOREST *The health of forests is affected by many factors – some natural, some man-made. Telling them apart is often not easy.*

overwhelming effect, and rabbit numbers plummeted. There were, however, some unforeseen side effects: without rabbits the remaining grassland quickly became overgrown, and the thyme and ants disappeared. With them went the large blue.

GLOBAL HABITAT CHANGE

There are about 9000 species of birds alive today and at least 100 000 species of butterflies and moths. The total number of living things that have been identified so far stands at about 2 million, and many thousands of species are added each year. In cataloguing this tremendous diversity of life, biologists are like astronomers scanning the skies for stars: the harder they look, the more they find. Until recently the total number of roundworms or nematodes known to science stood at about 15 000 species. However, investigations of sediment on the seafloor have shown that this figure is probably just a tiny fraction of the true total, which some biologists guess to be more than a million. What is true of nematodes is probably true of many other simple forms of animal life.

Seen against these figures, the retreat of a single butterfly from part of its former range may not seem like an event of global importance, and even the loss of a single species of woodpecker – albeit one of the world's largest – may not sound too grave. But the ivory-billed woodpecker is not an isolated example, and neither is the large blue. Habitat change has already claimed many other species, and has reduced thousands more – from tigers to macaws – to tiny fragments of their former range. Almost all of these animals are either large or conspicuous, which is why we know so much about them. Many other living things are too small to attract this kind of attention, and for these species scientists can only guess at the overall impact of habitat change.

The implications of this pruning of life's diversity are difficult to predict, because the natural world is too vast and complex to be modelled by any computer. At one extreme some scientists take the view that the Earth has such a variety of living things that the loss of some species will have little effect on those that remain. According to this view, we could survive quite well without many of the million or so species of nematode worms that may currently exist, or without some of the 300 000 species of flowering plants that botanists have so far identified. The living world would certainly be less diverse, but – so this argument runs – it would continue to function perfectly well.

For other scientists, including many of the world's leading ecologists, the man-made loss of biodiversity is a cause for considerable concern. This concern stems partly from the fact that the loss of any species deprives the world of a unique collection of genes, which are the chemical instructions responsible for assembling living things and controlling the way they work. Some of these genes create characteristics such as disease resistance, the ability to synthesise natural pesticides, or the ability to grow high-yielding crops, which are all of great potential value to us. When a species dies out, a pool of genes is lost and cannot be brought back into existence.

On a broader stage, the sheer diversity of life may also confer benefits of a quite different kind. One of the easiest ways to demonstrate this is to create an ecosystem of the most basic type. The ingredients are simple: a small tank of water and two types of microscopic organism, one of which lives by harnessing sunlight and the other by preying upon the first. As time goes by the numbers of the two creatures often oscillate wildly as the predators first feast on their prey and then begin to suffer as the numbers of prey begin to tumble. When the predators eventually start to starve, the prey often recovers, and the cycle may begin again. But the miniature ecosystem rarely achieves a stable state; instead, its inhabitants lurch from one extreme to another.

If a greater variety of organisms is added to the tank, something interesting often happens. The interactions between the tank's inhabitants now become more complex, making it difficult for any one species to become dominant. As a result the population swings become more subdued, and periods of relative stability set in. In this

CAPTIVITY *Born in the safety of San Diego zoo, this young black rhino can look forward to a secure if uneventful life. Its wild relatives are less fortunate.*

kind of environment no single species has the upper hand, but none runs the risk of eating its way to starvation.

Rudimentary though it is, this benchtop experiment may explain why diversity is so important for life. The living world provides many examples of sudden population oscillations – such as those of marine fish – but these may be just a hint of what could occur if there were no natural cushioning against their peaks and troughs. Many ecologists believe that this cushioning is brought about by the complex interactions between huge

numbers of species, and that we interfere with these interactions at our peril.

Much has changed in the world since September 1914, when the last passenger pigeon reached the end of its unusually long life. Areas that were then remote wildernesses – such as wetlands and tropical rain forests – have been opened up and converted into farmland, and in many parts of the world natural habitats have been replaced by much simpler ecosystems in which a few species – principally food plants, weeds and pests – exist to the exclusion of most others. Our impact on nature has never been so great, but nor has our interest in trying to protect the Earth's natural inhabitants against man-made change.

During this period the natural world has suffered many setbacks, but there have also

been some notable successes. The closing years of the 20th century have seen important moves in combating the threat to wildlife through pollution and climate change. National parks and wildlife reserves have been set up in many countries, and an entire continent – the Antarctic – has been made safe from development. Individual species have also been helped back into a healthy state. The grey whale, which numbered just a few hundred animals at the turn of the 20th century, was given legal protection in 1946. Since then the numbers in the eastern Pacific have climbed to over 20 000, probably as many as existed there before whaling began in the 1840s. In southern Britain the otter is making a comeback after vanishing from most of the lowland rivers in the country, while in France the griffon vulture (*Gyps fulvus*) – a huge scavenging bird with a wingspan of over 8 ft (2.4 m) – is once again soaring over the steep mountain gorges that were its stronghold at the beginning of the century. For these animals, at least, the future looks encouraging.

For other species, including some of the world's largest land mammals, the prospects are much less certain. Conserving an animal like the grey whale has a relatively minor impact on human livelihoods, but conserving an animal like the tiger or the black rhinoceros presents problems of a much more intractable kind. Despite adapting successfully to habitats as diverse as the jungles of Indonesia and the coniferous forests of Siberia, the tiger cannot survive without immense amounts of space, and in an increasingly crowded world that space is fast disappearing. Hunting is the major threat for the black rhinoceros. For the sake of its horns alone the population of this magnificent animal, which can weigh over a ton, has been reduced from about 65 000 animals in 1970 to just over 2000 today. Many conservationists fear that the species is fast reaching the point where it will cease to exist in the wild.

Animals like the tiger and black rhinoceros existed on Earth long before humans learned to hunt, and then to farm, and began to transform their surroundings. Yet time is no longer on their side.

CHARTING THE FUTURE

4

COLOURED EARTH *Brilliant orange waste from a lead mine dries and cracks in the sun.*

SINCE THE DAWN OF RECORDED HISTORY, PEOPLE HAVE ALWAYS WONDERED WHAT THE FUTURE WILL BRING. SOME HAVE LOOKED FORWARD WITH OPTIMISM, WHILE OTHERS HAVE SEEN GROUNDS FOR ALARM. TODAY THE SAME ATTITUDES CAN BE FOUND, BUT PREDICTING THE FUTURE HAS CHANGED. INSTEAD OF BEING AN ART SHROUDED IN MYSTERY, IT HAS BECOME A SCIENCE BASED ON FACT. WE CAN GAIN SOME IDEA OF THE FUTURE BY LOOKING AT EVENTS THAT HAVE HAPPENED IN THE PAST AND GAUGING THE LIKELIHOOD THAT THEY WILL HAPPEN AGAIN. BUT THE FUTURE IS NOT SIMPLY A REPETITION OF THE PAST. BY ASSESSING TECHNOLOGY'S MOST RECENT DEVELOPMENTS, FUTUROLOGISTS TRY TO ENVISAGE THE VARIOUS CHANGES THAT LIE AHEAD.

SERRIED RANKS *Tomorrow's cars await delivery.*

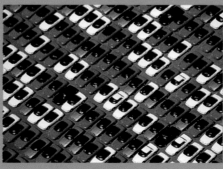

EARTH IN THE 21ST CENTURY

What will the Earth be like as we move into the 21st century?

Nobody truly knows, but there is no shortage of expert opinion. Some foresee a global ecological crisis; others are confident that we can moderate our impact on the world.

In 1902, at the dawn of a new century, the English writer H.G. Wells published a book containing some of his thoughts on the coming 100 years. It was written at a time of great change, particularly in the fields of physics and engineering; the automobile had recently been invented, the first steerable balloons had taken to the skies, and a new and mysterious source of energy – radioactivity – had been discovered. In *Anticipations* Wells speculated on the way in which scientific progress was likely to affect everyday life.

Transport was one of his favourite themes. He correctly predicted the impact of steam trains in allowing cities to expand, but unlike most people of his time he realised that steam locomotives had their limits. With their huge weight, Wells wrote, these mechanical giants would never be able to run on ordinary roads, and this was their major weakness. 'People of today take railways for granted as they take the sea and sky; they were born in a railway world, and they expect to die in one.' But were steam trains likely to remain the predominant method of travel on land? Wells, for one, doubted it.

What would oust them, he wrote, was something more versatile: a 'highly mobile conveyance capable of travelling easily and swiftly to any desired point', and – unlike a train – of 'traversing, at a reasonably controlled pace, the ordinary roads and streets'. He envisaged a 'motor carriage … capable of a day's journey of 300 miles [480 km] or more' and prophesied that, in the not too distant future, people would be able to travel in comfort at speeds above 70 mph (112 km/h). At a time when the first automobiles chugged and spluttered along at a few miles an hour, it was a remarkable statement.

Wells went on to consider how this kind of transport might affect the landscape. The roads of the future, he wrote, would be surfaced with asphalt and, because of the speeds involved, the traffic in both directions would have to be strictly separated. He saw it as an exciting prospect. 'Through the varied country the new wide roads will run, here cutting through a crest and there running like some colossal aqueduct across a valley, swarming always with a multitudinous traffic of bright, swift (and not necessarily ugly) mechanisms . . .' As a vision of today's highways and the cars that run on them, it was extraordinarily accurate.

A NEW KIND OF GAS *A car fills up with liquid hydrogen at a solar-powered filling station. The Sun's energy is used to split water into hydrogen and oxygen. Burning the hydrogen as a fuel recombines it with oxygen, forming water as a harmless waste product. A maglev monorail (opposite) uses magnetism to float along its track. The result is an efficient, friction-free ride.*

However, despite his grasp of the scientific advances of his time, even Wells was fallible. At the end of the 19th century the race was on to design a powered aeroplane, and a reader asked Wells what the prospects were for this kind of craft. Wells was dismissive. There was no doubt, he said in a brief footnote, that powered flight would eventually be achieved (it actually happened in 1903), but he had little faith in its lasting value. 'I do not think it at all probable,' the great visionary wrote, 'that aeronautics will ever come into play as a serious modification of transport and communication.'

This remarkable slip from such a perceptive thinker highlights the hit-and-miss nature of looking into the future. Because most predictions are based on current trends, the unexpected often throws them wildly off course. Wells was normally quite good at skirting around this trap, but many of his contemporaries fell headlong into it. For example, at the time *Anticipations* was published there was growing concern that great cities would eventually become choked to a standstill – by the dung of innumerable horses. It seems laughable today, but at the time it was deadly serious.

CATASTROPHISTS AND CORNUCOPIANS

Despite dangers and pitfalls such as the horse-dung 'problem', futurology is just as much alive as it ever was. So what do today's scientists think the Earth will be like as we move through the next 100 years: a planet at the mercy of a changing climate, tainted by pollution, and burdened by a growing human population, or one in which we manage to overcome many of the problems that now face us?

In a field as complex as this there are countless shades of opinion. Many scientists take the view that, as in the past, the news is bound to be mixed. The future will turn out to be better in many ways, thanks largely to human ingenuity, but unexpected problems will inevitably occur, along with some unexpected solutions. On each side of this centre ground, however, are two groups of experts who hold more outspoken views. The members of one camp, often dubbed the doomsters or catastrophists, see a world in which a soaring human population will inevitably make almost everything worse.

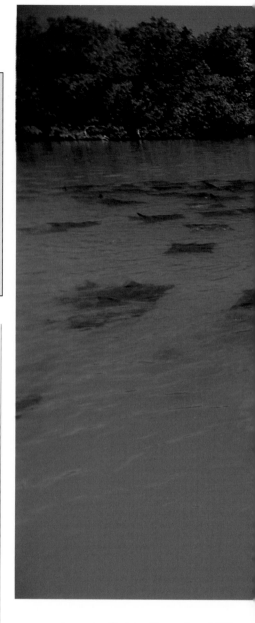

LESSONS FROM LONG AGO

In 1968 the US ecologist Garrett Hardin published an essay describing the fate of common land in medieval Europe. Because this land belonged to everyone, it was in each person's immediate interest to graze as many animals on it as possible. However, the land became degraded, and everyone lost out in the long term. In tomorrow's world there will be a growing conflict between the long-term need to nurture resources and the short-term need to use them.

The others, known as the cornucopians, turn this idea on its head. They think of people not as a problem but as an asset, and believe that an expanding population will spur on the technology that makes the world a more hospitable place.

In 1980 two leading adherents of these contrasting viewpoints put their beliefs to the test. Paul R. Ehrlich, an American ecologist who first coined the phrase 'population bomb', agreed to a $1000 bet with Julian L. Simon, an economist. Simon had issued a challenge to the catastrophist camp, asking them to select any five commodities and wager on their future price at the end of a ten-year period. If population growth really did make resources scarcer, the prices should rise, but if – as Simon believed – technological improvements made essential resources easier to find and to produce, the prices should fall. Ehrlich chose five metals that are widely used in industry, and the bet was drawn up.

In the years between 1980 and 1990 the world's population rose by several hundred million, which meant that resources were being used at an

PROBLEM OR SOLUTION?
A crowded Brazilian beach highlights the effects of population growth. Many see this growth as a problem, but some view it as a spur to beneficial change.

ever greater rate. But in November 1990, when the time to settle the bet arrived, the prices of all five metals had fallen. The change was as Simon had predicted, and the cornucopian camp had won.

Conclusive though the result seemed at the time, the bet did not bring the debate to an end. Today's catastrophists still argue that, in the long term, future trends will be negative, while the cornucopians are equally confident that they will not be. To complicate matters, the two sides are also beginning to focus on different trends. For the cornucopians the important ones are those that directly affect human welfare, such as life expectancy and the price of food. For the catastrophists the key trends are ones that affect the whole of the living world, such as the extent of tropical forests

ALGAL BLOOM *In this high-altitude photograph, blankets of algae appear as faint streaks off the coast of Australia.*

and changes in global temperature. In years to come these trends are likely to move in different directions, making it even harder for people to agree whether the world is getting better or worse.

ENGINEERING THE WORLD

At the beginning of the 20th century, when H.G. Wells foresaw the shortcomings of steam, bold engineering ventures gripped the public imagination. Eager crowds turned out to witness the launch of larger and faster ocean-going liners, and gasped with awe at the inaugural flights of immense streamlined airships, which were no longer simply gas-filled bags blown about at the mercy of the wind. New bridges were built across obstacles that were previously thought to be unbridgeable.

SEASONAL GATHERING *Like almost all forms of life at sea, these cow rays ultimately depend on microscopic algae. In the future, humans may control the levels of these tiny but vital organisms.*

During the new century the achievements multiplied. First came the perfection of the internal combustion engine and the conquest of the air, and then the exploitation of atomic power and our first journeys into space. But while heavy engineering will undoubtedly still play an important part in the next century, some scientists think that we may be on the brink of a technical revolution of a quite different kind. Instead of creating vehicles or static structures, this

new form of engineering deals with the whole of the Earth.

In 1995 a small group of boats made their way to an area of the Pacific Ocean just west of the Galápagos Islands, to give one of these engineering projects its first outdoor test. Despite the fact that the Galápagos are on the Equator, the islands are bathed by the distant reaches of the Humboldt Current, which flows up the coast of South America from Antarctica, and the water around them is deep, cold and clear. Like many regions washed by cold currents, this area is rich in mineral nutrients, and contains a large amount of dissolved oxygen. However, for reasons that have until recently been a mystery, the deep water off the Galápagos Islands is the marine equivalent of a desert, with remarkably little life.

As the boats tracked backwards and forwards across an area of about 25 sq miles (65 km²), they scattered an unusual cargo – ferrous sulphate, an iron-containing compound that quickly dissolves in water. Having unloaded over half a ton of the yellow-brown powder, the boats hove to, and their occupants waited and watched.

They did not have to wait long, and the result turned out to be far more impressive

THE FUTURE FOR NUCLEAR POWER

In the 1950s nuclear power was heralded as the energy source of the future, and one that would eventually make electricity 'too cheap to meter'. Today nuclear power has a quite different reputation. A succession of accidents, including the Chernobyl explosion of 1986, have made many countries wary of this source of energy and the problems it brings.

Today's nuclear reactors work by fission, which involves splitting atoms and releasing the energy that binds them together. The fuels used are the elements uranium and plutonium, whose heavy atoms are unstable and susceptible to being broken apart. The atoms are split by bombarding them with neutrons, and as they break up they release more neutrons, so a chain reaction begins. The reaction generates immense amounts of heat, but it also produces potentially deadly gamma-ray radiation as well as radioactive waste.

CHAIN REACTION *When an atom of uranium is bombarded with neutrons the atom splits, producing more neutrons and releasing a vast amount of energy.*

However, fission is not the only way of releasing nuclear energy, and physicists are hopeful that the alternative method – fusion – may eventually become a practicable and much safer source of power. Unlike fission, fusion uses hydrogen, which is the lightest and most abundant of the elements, and it generates only tiny amounts of radioactive waste. Unfortunately there is a problem: to make the atoms fuse, the hydrogen fuel has to be heated to about 100 million°C (180 million°F) – a temperature that would instantly

vaporise any container. Several experimental reactors have been designed to bypass this apparently insuperable problem. In one, known as a torus or tokamak, the fuel is confined in a hollow doughnut-shaped chamber by an intense magnetic field, while in another, called a NOVA laser, it is compressed by beams of light. Both these experimental systems have successfully triggered fusion, but only for fractions of a second, and so far they use up more energy than they release.

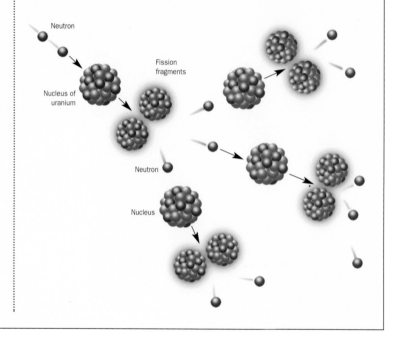

Neutron

Nucleus of uranium

Fission fragments

Neutron

Nucleus

the whole of the atmosphere. By doing this it would help to combat the greenhouse effect, which is currently increasing surface temperatures all over the globe.

The Galápagos experiment showed that such blooms can be triggered, but so far as the greenhouse effect is concerned the gain from this particular bloom was short-lived; the algae eventually died and decayed, so their carbon was released back into the water and the air above. But there are places – particularly in the Southern Ocean – where the surface waters are constantly sucked downwards, at the start of a seafloor journey that keeps them far away from the distant air. If one of these areas were constantly 'fertilised' with iron, the algae would absorb carbon dioxide from the air, but almost as fast as they formed they would be swept away by the marine equivalent of a conveyor belt, to be locked up in sediment in the depths below. With the algae buried in the seabed, their carbon would not be able to return to the atmosphere in the normal way.

than anyone had expected. Within a few days the normally transparent surface waters turned bright green, forming a gigantic stain that looked like an immense blob of ink. The blob drifted westward in the current and remained visible for several weeks, until it eventually began to fade away.

This green stain was produced by a sudden burst of biological activity. Marine biologists knew that in this part of the tropics, planktonic algae – microscopic plant-like organisms that drift near the surface – were unusually scarce, despite having all the sunshine they could possibly need. They concluded that the algae must be short of some essential nutrient, and pinpointed iron as the most likely possibility.

As the experiment showed, their analysis was correct. When iron was added to the water, the algae underwent a 'bloom' or population explosion and the formerly unproductive surface waters erupted into life.

Algae form the basis of almost all marine food chains, so the more algae there are the more life the oceans can support. But the experiment had a more subtle aim than simply 'greening' the seas. When algae grow they absorb carbon dioxide from the water, and the water in turn absorbs carbon dioxide from the air. A giant bloom of algae – many times the size of the one produced by the Galápagos experiment – could theoretically soak up so much carbon dioxide that it could alter the gas's concentration in

This opens up the possibility of some truly far-reaching adjustments to the Earth's atmosphere. According to some calculations, in 100 years this relatively simple procedure could reduce carbon dioxide to pre-industrial levels, and the greenhouse effect would be a thing of the past.

EXPERIMENTING WITH THE EARTH

Geoengineering has already generated a clutch of remarkable ideas. As well as plans to fertilise the seas there are schemes to reverse the flow of rivers, to tow icebergs to the tropics to provide fresh water, and to generate power by connecting landlocked depressions to the open sea. Some of these projects have reached an advanced state of planning, and rely on technology that is already well within our grasp. However, as yet, few have been put into action. One reason for this delay is that schemes of this size are often prohibitively expensive. Another and even more intractable problem is that, like the future itself, the results of this kind of engineering may not be quite as predictable as they seem.

The Galápagos experiment illustrates this problem well. On the face of it, triggering algal blooms would be a superbly effective way of tackling one of our most pressing problems. Iron is easy to obtain and entirely safe to handle, and, because it makes up about 5 per cent of the Earth's crust, there is little danger of it running out. As well as lowering carbon dioxide levels, algal blooms would also increase the productivity of the seas. In time this would boost flagging stocks of fish, and generate much-needed supplies of food.

But there is a catch. Although iron is abundant, it is locked up in the form of iron ore. This ore would have to be mined, and the iron would then have to be converted into a soluble form. Once this had been done, it would have to be carried to the coast, loaded onto ships and taken out onto the ocean, and then dispensed in the correct place in precisely measured amounts. All this would consume energy, which would almost inevitably come from burning fossil fuels. Ironically, those fuels would release carbon dioxide – the very substance that the project would be designed to reduce.

A second question mark hovers over the biological effects of this bold restructuring of the oceans' natural cycles. Giant algal blooms may prove to be beneficial to other forms of life, but there is a significant

danger that they might be harmful. Some species of algae produce extremely potent poisons that kill fish, seals and other sea animals. If these algae established themselves in the iron-rich water they would be difficult to control, and the outcome might be an environmental disaster.

As yet no one knows how real any of these dangers are, so fertilising the seas, like many other geoengineering projects, remains an exciting but unproven possibility. A hundred years ago our forebears pressed on unhesitatingly in the pursuit of technological progress, but in the ecologically aware 21st century caution is likely to be the watchword.

POWER FROM THE SEA

While geoengineering may have to struggle to prove its worth, other kinds of technology are certain to have an increasing impact on the Earth in years to come. Mirrors are already helping to gather heat – by reflecting sunlight onto a central point where the heat can be used – and to convert it into electricity. During the 21st century they are certain to be joined by some unusual-looking machines as engineers experiment with different ways of producing non-polluting power from the wind and the waves.

If construction goes according to plan, some of these machines may become familiar sights off some of the world's coasts. One design, currently being tested off the stormy north coast of Scotland, squats on the seabed in about 50 ft (15 m) of water. While the waves pummel passing ships and the wind buoys up gulls in their effortlessly wheeling flight, the Osprey gathers energy from both and sets about putting it to work. It funnels waves into a collector chamber, where the moving water blasts air through a set of turbines to generate electricity. Another turbine, mounted above the waterline, harnesses the wind overhead.

On the other side of the globe, in Japan, marine engineers have come up with a different kind of wave-powered generator. Measuring about 160 ft (50 m) long and 100 ft (30 m) wide, the Mighty Whale is soon to be the largest of a new generation of floating power plants. Despite its name, this machine looks less like a single whale than several side by side. It has a hollow 'head' and fin-like 'tails', and it automatically turns into the wind to swallow oncoming waves. Like the Osprey, the Mighty Whale uses the force of waves to compress air, but instead of staying in one place it can be moved wherever it is needed.

WAVES AT WORK *Guided by a pair of tugs, the Osprey wave-powered generator heads out to sea. Machines like this produce power without any pollution.*

The Osprey and the Mighty Whale represent new uses for old technology – the turbine, for example, has existed in various forms for hundreds of years. But while these machines garner power from the waves, a much newer form of technology will help 21st-century humans to plan the best way of using the Earth's limited resources. That technology is based on microelectronics, and it will create an explosion of knowledge about our planet, dwarfing what we already know today.

EYES IN SPACE

From the moment that people first ventured aloft in hot-air balloons, flight has given us a completely new perspective on the surface of the Earth. Structures that are too big to be studied on the ground can be taken in at a single glance, and features that were formerly hidden – from the boundaries of former fields to entire ruined cities of ancient civilisations – often become visible for the first time. Early balloonists and pilots recorded these details with sketch maps and later with hand-held cameras, but with the advent of microelectronics and satellites this kind of portraiture has become vastly more detailed.

Satellite technology already allows us to make some remarkably precise measurements of the Earth's surface features. For example, weather satellites give meteorologists instant information on the cloud cover all over the globe and enable them to predict the paths of hurricanes and typhoons. Satellites also monitor the average height of the oceans, supplying measurements accurate to within a fraction of an inch, and chart the movement of entire continents, which at its fastest amounts to no more than 7 in (18 cm) a year. By beaming electronic signals down from space they form the basis of global positioning systems, which allow anyone with a handheld receiver to fix his or her position on the surface to within a few feet. To the navigators of earlier centuries, who struggled to achieve the same ends with cumbersome sextants and unreliable chronometers aboard wave-rocked ships, this alone would have seemed a miraculous achievement.

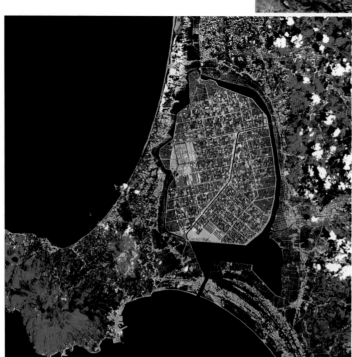

ORBITING EYE *A high-definition Landsat image shows reclaimed land off the Japanese island of Honshu. Vegetation is shown in red, and bare ground in blue and grey.*

Some satellites hang high above particular points on the Equator, matching the planet's rotation and keeping an entire hemisphere under their steady gaze, while others follow orbits that loop over the poles, so they gradually cover the whole of the Earth's surface as the planet turns beneath them. These satellites orbit at relatively low altitudes and can produce detailed images of the surface and its vegetation. Their pictures have revealed the inexorable shrinkage of the tropical rain forests, and the haze produced by burning timber. By viewing the same areas in light of several different wavelengths, including visible colours and infrared, satellites can distinguish between natural vegetation and crops, and can reveal man-made surfaces such as roads and roofs, which reflect more heat than their surroundings. In the coming century the prospects for these eyes in the sky seem almost infinite. Satellites will map the world with ever-increasing clarity, and

the information they gather will be analysed with greater and greater precision.

As a pointer towards things to come, an American satellite launched in 1996 will photograph the Earth's surface in more than 380 different wavelengths of light, which is many more than earlier satellites

AFRICA FROM SPACE *This satellite image takes in the volcanic craters of northern Tanzania and part of the Serengeti National Park. The white area is a salt-laden lake.*

One of the great strengths of this kind of technology is that satellites move across accessible and remote places with equal ease. In the next century satellites will peer down into the surface of the seas for signs of pollution and scan forests, swamps and scrubland, automatically identifying the exact places where disease-carrying organisms, such as malaria mosquitoes, are likely to breed. When man has this kind of pinpoint information, the fight against these two threats is likely to become both easier and more efficient.

LIFE RESHAPED

In the early 20th century the first microchips were still 75 years away, and as a result it is hardly surprising that writers such as H.G. Wells saw the future largely in mechanical terms. But if Wells and his contemporaries could be brought forward to the present, microelectronics would not be the only technology that would fire their imagination. At the threshold of the 21st century a completely new branch of science seems poised to shape our future. Instead of engineering machinery, this science engineers living things.

designed to image the Earth. The satellite will orbit the planet at a height of just 320 miles (515 km) and will be able to pick out objects that are just 16 ft (5 m) across. Because different minerals and plants reflect separate wavelengths in differing degrees, computers processing the satellite's data will be able to determine what the space-borne cameras have picked up in every frame. They will be able to distinguish between mineral types, and once the appropriate computer software has been perfected they may even be able to recognise individual species of plants.

From the days when people first bred crops and tamed animals, humans have accidentally or deliberately interfered with the characteristics of other forms of life. Initially it was a fairly haphazard process, and consisted simply of giving some species a head

continued on page 146

Genetic Engineering

In early 1997 scientists working in Scotland announced an astonishing event. For the first time they had successfully 'cloned' a mammal, using a single cell from an adult ewe to produce a healthy female lamb. Because the lamb was cloned it had exactly the same genes as its mother. Instead of being a normal daughter, it was more like its mother's identical twin.

The experiment was remarkable in many ways. One was that, unlike earlier attempts at cloning, it did not use an egg cell, which is a cell already destined to become a new animal. Instead it used an ordinary cell from the mother's body. The cell's programming was reset, and once it was implanted in the sheep's womb it began to develop like a fertilised egg. The cell came from a female donor, but it could have equally well come from a male. A male cell would still need to develop inside a female, but genetically it would produce a lamb that had a father but no true mother.

To many scientists, experiments like this represent a huge achievement and are one

WEEDED OUT *This dish contains tobacco seedlings that have been treated with a herbicide. The normal seedlings have died, but thanks to an extra gene the engineered ones have survived.*

of the most promising ways of increasing the world's food supplies and fighting disease. Instead of reshuffling genes that species already possess, using slow and often difficult breeding techniques, tomorrow's scientists will transfer useful genes between species. With cloning they will then be able to copy the results as often as required. However, for many people in the world outside – and some scientists – these benefits come at too high a price. Like the fabled world of Dr Frankenstein, who created an uncontrollable

RESULTS ON SHOW *Agricultural researchers examine experimental fruit trees that have been engineered to produce more blossom.*

monster, genetic engineering seems like a technology running out of control.

At the heart of this debate are some remote but alarming possibilities. Foremost among them is the threat that an engineered microorganism might somehow escape from the restrictions of the laboratory or test-plot and reach the wider world beyond. Here it would be beyond our control, and it might cause disease or pass on its genes in an unpredictable way.

Microorganisms certainly have been known to escape. In 1995 a virus broke out from a test area on Wardang Island, near Adelaide in Australia. The virus was being evaluated for use against Australia's introduced rabbits, but it was not due to be used until it had proved harmless to native wildlife. Although it was meant to be carefully confined it managed to reach the mainland about 3 miles (5 km) away, and once ashore it spread as rapidly as a bush fire. Within six weeks it had spread over 200 miles (300 km), leaving millions of rabbits dead. This virus was not genetically engineered, but its escape – opponents say – highlights the dangers involved in working with ones that are.

Proponents of genetic engineering point out that it is very unlikely that a genetically engineered microbe could infect anything other than its normal host. However, genetic engineering poses subtler problems than

CLONED PLANTS *Cloning is much easier with plants than with animals. These cloned plants are growing on a special sterile jelly.*

deadly epidemics. One of these is interbreeding. Many crop plants are currently being engineered to give them resistance to disease or to selected herbicides. It is possible that these plants might breed with their wild relatives, thus passing on their engineered genes. If this happened it could create new breeds of 'superweeds' that would be difficult to control.

Supporters of genetic engineering are confident that these dangers can be avoided. Even so, widespread cloning raises further questions that will take longer to resolve. In recent decades 'traditional' breeding techniques have produced a narrow range of plant and animal varieties that we depend on for our food. As a result the genetic diversity of these plants and animals has already been greatly reduced. If cloning is widely adopted this genetic narrowing will be accelerated even further, because cloning cuts out the mixing of genes that happens when living things reproduce. The result is a potentially dangerous situation in which we come to rely on a small collection of genetically identical sources of food. If one cloned plant or animal becomes infected by a disease, all its cloned relatives will be equally at risk.

Genetic engineering has been carried out since the late 1970s, and so far it has produced no disasters and many benefits. However, during the 21st century its impact is likely to be far greater. For this reason it is under intense scrutiny, as scientists and non-scientists try to ensure that it maintains its current safe record.

PRODUCTION-LINE PLANTS *Each beaker contains tissue that will grow into a plant. Unlike normal seedlings, the plants will be genetically identical.*

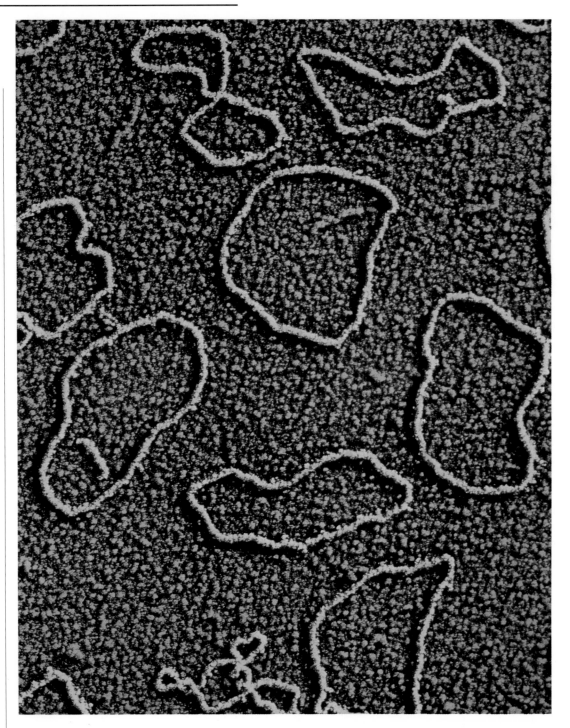

CHEMICAL CARRIERS *Plasmids, or loops of DNA from bacteria, are often used to ferry new genes into cells. Plasmids can copy themelves and any new genes inserted into them.*

heredity, there were limits to his understanding. Something obviously carried the instructions for different characteristics when living things bred, but Mendel had no idea what that something might be.

Mendel's work went largely unnoticed in his own lifetime, and nearly four decades passed before its significance became apparent. In 1902 an American biologist called Walter Sutton was studying chromosomes and puzzling over the function of these microscopic thread-like structures that lie at the heart of every living cell. Chromosomes had been known about since the 1880s, when new dyes made them visible for the first time, but Sutton made the crucial mental leap which led to our understanding of what they actually do. These tiny threads are inherited in paired sets – one set coming from each parent – making them just the sort of vehicles, Sutton reasoned, that are needed to carry instructions when living things breed. With this realisation the focus of research moved from Mendel's outdoor world, with its pea plants, pollen brushes and notebooks, into the very different world of individual cells and the information-carrying chemicals that are locked up within them. The science of molecular biology was born.

With the discovery of the structure of DNA, in 1953, the new science came of age. This breakthrough revealed that all forms of life on Earth – from an amoeba toiling through the film of water around a single speck of soil to a male bird of paradise

start over their competitors, for example by planting their seeds or by protecting them from predators. From here the next step was often to select the most useful varieties of each species and to control the way they bred. Normally this would involve plants or animals belonging to the same species, but in some cases parents from different species could be persuaded to breed, creating hybrids that rarely occurred in nature. These hybrids, which included animals such as the mule (half donkey, half horse), were not always able to breed themselves, but their useful

qualities, such as the mule's sure-footedness, often meant that they were highly valued.

In the 1860s Gregor Mendel, an Austrian monk, carried out thousands of painstaking experiments in an effort to find out exactly how characteristics are handed on when living things breed. Using pea plants, he made the key discovery that characteristics do not simply blend together when they are passed on; instead, the versions from each parent are kept separate. They both remain intact, even though only one is outwardly shown. But while Mendel grasped the principles of

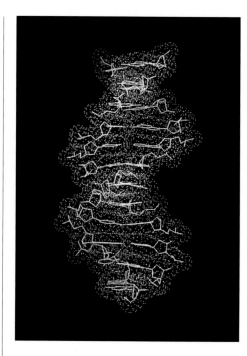

INSTRUCTIONS FOR LIFE
This computer-generated image shows a tiny fragment of DNA.

displaying its plumage on the floor of a New Guinea rain forest – owe their characteristics to an almost endless series of just four chemical letters, arranged in precise but varying sequences to form molecules of DNA. In many living things, including ourselves, up to 95 per cent of the sequence seems to be meaningless – like random scribblings on an enormous notepad – but buried between these scribblings are genes, or lists of instructions that shape living things and organise the way in which they work.

CHANGING THE CODE

Once molecular biologists learned to read these coded sequences, the next step was to alter those sequences. From the 1970s onwards chemicals called restriction enzymes have been used to cut up DNA molecules at specific sites, creating isolated genes that can then be combined in new ways or even inserted into quite different organisms. In the space of a few decades we have acquired an awesome ability to change the living world.

In nature, DNA is altered all the time. It can be disrupted by the ultraviolet light in strong sunshine, by chemicals in the environment, by mistakes made when cells divide, and also by molecular wear and tear. However, few of these alterations survive for long because they are quickly repaired. Teams of chemicals shuttle up and down the DNA strand, checking for signs of damage, and when they find an incorrect letter they cut it out and insert a replacement. Sometimes a mistake is too great to be patched up in this way, and it becomes a permanent feature of the DNA strand. In most cases these permanent accidents, which are known as mutations, are either unhelpful or even dangerous, but just occasionally – perhaps less often than one instance in a million – they produce a new and more useful version of a gene. If this new version of the gene is passed on to its owner's offspring it may start to spread.

This process is the driving force behind evolution, because it produces the tiny variations that allow species to change as time goes by. But while mutations occur all the time, it can take hundreds or thousands of generations for a gene to become widespread. Even when it has reached this stage it is normally trapped within a single species. It can travel vertically from one generation to the next, but except in rare cases it cannot travel horizontally, to cross the invisible barriers that separate one species from another.

These are the established rules of nature, which have operated for at least 3.8 billion years. But with the advent of genetic engineering, a previously inconceivable world has opened up. Using restriction enzymes as chemical scissors, genetic engineers can remove individual genes from one organism's cells and insert them into the cells of another species. Because DNA has the same underlying structure whether it comes from a bacterium or an elephant, a transplanted gene will slot into place, and the host cell will put its instructions into effect.

This transplanting of foreign genes can produce some bizarre results. For example, plant breeders have managed to insert genes from fireflies into tobacco plants, making them glow with a firefly's greenish-yellow light. This might sound like an absurd achievement, but it has a practical aim. When genes are moved from one organism to another it is not always easy to confirm that they are present and fully functioning. But if the genes include a visible marker – for example, a gene that makes the recipient light up – success is easy to spot.

So far genetic engineering has had a relatively minor impact on us and the world around us. Genetically engineered bacteria produce some valuable substances, such as insulin and human growth hormone, and genetically engineered food has started to appear on the shelves of some supermarkets. Genes that kill insect pests have been successfully transferred from bacteria to crops such as tomatoes, and other kinds of disease resistance have been transferred from plant to plant. But as we move into the 21st century, and laboratory techniques become more refined, there is little doubt that genetic engineering will play an increasing part in daily life.

BUBBLING UP *A block of emulsion agar gel – the world's lightest solid substance – sits on a cushion of soap bubbles. In the future, genetic engineering may produce many new, unusual materials.*

EARTH'S DISTANT FUTURE

Today's Earth is a favourable place for living things, but in the distant future conditions may abruptly change, and we may not survive. In this far-off world, new forms of life will evolve in the place of ones that have disappeared.

Imagine the world 65 million years from now – as far into the future as the dinosaurs are in the past. Humans have long since disappeared, and some major changes have taken place on the surface of the Earth. The Atlantic Ocean has widened and the Pacific has shrunk, Australia has collided with Indonesia, a slice of California has reached Alaska, and Africa's Great Rift Valley has opened up and been drowned by the sea, creating a gigantic island surrounded by the Indian Ocean. Because the continents have shifted, warm water now flows northwards to reach the Arctic. The sea ice has melted, and, more importantly, the ice cap on Greenland has also largely vanished, raising sea levels by many feet.

As well as the Earth's topography, life has also changed. Mammals have largely disappeared, and in the warm and humid climate across many parts of the planet, insects are the most successful animals. The verdant landscape is dotted with their extraordinary nests, some almost

FALLING WALLS *In Kenya,*
ancient ruins show nature's
power to reclaim lost ground.

100 ft (30 m) high. In a world like this, which is so distant from our own, what evidence would show that we once existed, and that we dominated life on Earth?

The answer to this question is: surprisingly little. Despite the tremendous impact we have had on the planet, our tenure on it has been brief. Even our largest and most ancient structures date back just a few thousand years, which means that they have yet to be truly tested by time. Although we have been responsible for a sudden decline in life's diversity, life itself has an almost inexhaustible capacity to adapt and survive. Working hand in hand with the effects of wind, rain, frost and fire, living things would be quick to dismantle much of what we have created. Some signs of our former achievements would still exist, but they would be remarkably hard to find.

THE DISAPPEARING CITY

Modern cities are the most complex environments on Earth. Like living tissue they take in supplies from their surroundings, and they last because they are constantly maintained, repaired and rebuilt. Much of this work goes unnoticed, but if it suddenly came to a halt the effect would be profound.

Within five years plants would already have escaped from parks and gardens, and would be spreading through the streets. At first they would exploit existing fractures in asphalt and concrete, but gradually their dead leaves would build up, forming an organic layer over the surface – the beginnings of a fertile soil. Without humans to sweep it away this layer would eventually become deep enough to retain moisture all the year round, allowing tree seeds to germinate and take root. These roots would fan out

DESERT SURVIVOR

A remarkable desert plant from
Bolivia (left) has adapted to an
extreme environment. This kind
of tenaciousness ensures life's
long-term hold on our planet.

over the ground, seeking out the smallest crevices in the roads or pavements beneath, and they would also push their way between the joints of drainage pipes and reach down among foundations. The ground would become a living environment once again.

Meanwhile, above ground, buildings would be under attack. Within ten years peeling paint would be flaking away, and woodwork would begin to decay. Fire – an increasing hazard in an empty city – would soon consume anything flammable, and it would also shatter windows, sending shards of glass cascading into the streets below. The

heat produced by these fires would make steel girders twist and buckle, producing hairline cracks in their concrete cladding. Once these cracks had appeared, rain would be able to seep inside where it would rust the metal within. When steel corrodes, the rust it forms expands with an almost unstoppable force, which can make buildings shed their walls and floors and tear down bridges and flyovers. By the time two centuries had passed, all wooden and brick structures would be in an advanced state of decay; after three or four centuries steel-framed buildings would be collapsing as well.

Ironically, much older buildings would often fare better than this. The great cathedrals of Europe, built of solid stone, would soon lose their roofs, ornamental spires and decorative carvings, but their walls would prove more resilient. However, after centuries of exposure to the elements even these would crumble and fall. The last structures to disintegrate would be those built in dry places, such as the desert bordering the Nile. By the time Egypt's pyramids had been eaten away by wind-borne sand, most of the world's towns and cities

THE OLDEST STRUCTURES

The oldest known structure of any kind built by humans or our direct ancestors is believed to be a circle of stone blocks at Olduvai Gorge, Tanzania: it dates back about 1.8 million years. The oldest evidence of building is from a temporary campsite called Terra Amata, in southern France, which is about 400 000 years old. By contrast, the oldest intact or partly intact buildings date back just a few thousand years.

would be little more than overgrown irregularities in the landscape.

A million years from now, these ruins would show beyond any doubt that intelligent life had once dominated the Earth, but after tens of millions of years had passed the evidence would be harder to interpret. Many of the minerals in building materials would be leached away by rain or attacked by acids in the soil, reducing masonry, bricks and concrete to powdery fragments that would be crushed by the growing weight above. Changing sea levels would bury low-lying ruins under layers of sediment, and volcanic ash and lava would smother many of our traces on higher ground. In stabler parts of the Earth's crust some reminders of our former existence would survive, but after millions of years it would be increasingly difficult to prove that they were a legacy of intelligent life.

However, not all signs of human intelligence would perish in this way. Although the physical remains of our buildings and machines would be quite unrecognisable, they would still contain exceptional quantities of heavy metals, giving them a chemistry that would be as distinctive as a fingerprint. These geological anomalies would persist almost indefinitely, and to any future form of life able to detect them they would be remarkable features – particularly as they would be found in the same form in many different parts of the globe.

These metal signatures might be interpreted in a number of ways. Some bacteria accumulate unusual elements as part of their natural chemistry, and deposits of heavy

LEFTOVER LEAD *For future visitors to Earth, the remains of metal mining waste (left) might provide evidence of a vanished technology. By the time our radio signals reach the Large Magellanic Cloud (above) our species will be over 400 000 years old. Many species survive for less time than this.*

about 200 000 years it will have bridged the gulf to the Magellanic Clouds, which are our nearest galactic neighbours.

HARD TIMES

From a human perspective a planet without people seems a bleak prospect, even if it is a very distant one. But how likely is it to come about?

On this question the statistics of life are equivocal. Our species dates back perhaps 200 000 years, which in geological terms is only a short period. Some species have survived for far longer than this, but others have sprung up and then become extinct in shorter intervals of time. These short-term species are often ones that evolve in restricted habitats in small parts of the Earth's surface, and which have a highly specialised way of life. They can be successful as long as

metals can also be formed by collisions with objects from space. But the variety of metals in these seams would be highly unusual, making them act as pointers to the technologies of long ago.

A second piece of evidence would be even more persuasive but much harder to detect. Spreading out in an ever-expanding sphere, a front of radiowaves is already being beamed from the Earth into space, and will continue to travel even if our radio transmissions come to a halt. In about 26 000 years from now this front will have reached the edge of our own galaxy, and in

conditions remain stable, but any kind of change – such as the arrival of a competing species – can quickly prove disastrous. Humans are not like this. Compared with most forms of life we are remarkably unspecialised and adaptable, and we have a wide distribution. Seen in this light, the omens look good.

However, the Earth's history also shows that there have been times when swathes of species – including some that dominated life on land – have collectively died out.

The most recent of these mass extinctions brought the reign of the dinosaurs to an end about 65 million years ago, but a far greater extinction took place about 200 million years earlier. Known as the Permo-Triassic Crisis, it killed about 90 per cent of all animal species in the seas, along with two-thirds of the reptiles living on land. After this global catastrophe, life took millions of years to match the diversity it had had before.

Nobody yet knows what caused this greatest extinction of all, although recent research

suggests that it was a collection of different factors, including a drop in sea level, a period of intense volcanic activity, and some extreme lurches in the Earth's surface temperature. These changes probably took place over hundreds of thousands of years and hit living things like a succession of body blows, first destroying vital habitats, and then making conditions extremely difficult for the species that managed to survive. By comparison, the mass extinction that killed the dinosaurs seems to have been a single knockout punch, administered by a meteorite that struck the Earth's surface in what is now Central America. This impact would have created a shroud of dust that cut off the light from the Sun, producing a chaotic twilight world of dying plants and starving animals.

Despite the daunting scale of the Permo-Triassic Crisis, we are almost certainly better placed to ride out an event like that than a catastrophe of the kind that killed the

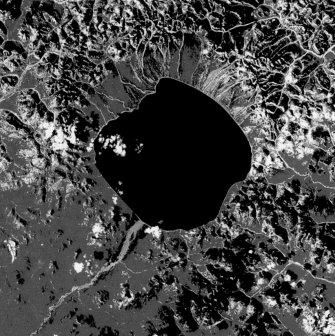

CLOSE MISS *A million-ton meteorite grazed the Earth's atmosphere on August 10, 1972, before continuing into space. This photo, taken in Wyoming, captures the trail that streaked across the sky. Left: Siberia's Elgygytgyn Lake – seen from a satellite – marks the site where a meteorite actually hit the Earth.*

dinosaurs. This is because we could probably develop the technology needed to survive long-term changes, even if life became harder than it is now. An abrupt global catastrophe would be a different matter, because our food resources are limited. If crops across the entire world were simultaneously devastated, as might happen after a major meteorite impact, the outlook would be dire.

The likelihood of this kind of disaster is impossible to calculate, but the laws of probability dictate that it steadily increases as time goes by. During the Earth's recent history we have experienced a number of impacts, such as the one that devastated part of central Siberia in 1908 and the one that created a fireball above the Pacific island of Tokelau in 1994. Objects of the size of the Tokelau meteorite, which weighed about 1000 tons, strike the Earth every few decades. Meteorites with the power to cause global destruction are much rarer, and arrive less than once in a million years, but while they are exceedingly uncommon the threat from them still exists.

Some astronomers and physicists believe that this is a threat that we could overcome. We already have the ability to detect meteorites that come close to Earth, and in the future these detection systems will become more refined, giving us a longer warning of their approach. A series of nuclear explosions in space, positioned to knock a meteorite from its original course, could avert a collision.

SUBTERRANEAN SURVIVORS

If the Earth does eventually experience another cataclysmic impact, or a replay of events like those that triggered the Permo-Triassic Crisis, the stage will be set for some remarkable biological changes. In a distant and hypothetical world, perhaps 200 million years from now, one thing would be

THE THREAT FROM EXPLODING STARS

In AD 1054 Chinese astronomers recorded the sudden appearance of a brilliant new star. It blazed so brightly that it could be seen in daylight, and though it soon faded it remained visible for two years. Five centuries later, in 1572, another of these celestial beacons flared up in a different part of the heavens, and the Danish astronomer Tycho Brahe described it in a book called *De Nova Stella*, meaning 'of the new star'.

Part of Brahe's title – the word *nova* – has passed into the language of astronomy. It is used to describe a star that collapses under its own gravity, suddenly brightening by more than 10 000 times to become visible to the naked eye. A supernova is a particularly massive example of this kind of celestial cataclysm; the novae witnessed by Brahe and the Chinese astronomers before him both fell into this category.

When it collapses, a supernova not only releases light but it also pours out energy in other forms, including cosmic rays. The Earth is bombarded by cosmic rays all the time, from the Sun and from outer space, but the atmosphere shields living things from their potentially lethal effects. However, if a supernova occurred relatively close to the Earth this shield would fail, and the planet's surface would be bombarded by a deadly barrage of subatomic particles.

According to recent research into supernovae, a deadly supernova within about 30 light years of the Earth can be expected about once every 250 million years. It sounds comfortingly remote, but it is still frequent enough to make it likely that supernovae have had a fateful influence in the history of life.

EXPLOSION *This photograph shows a disc-shaped galaxy, and just below it a brilliant supernova that flared up in 1972. It has since faded away.*

certain: the lowliest forms of life would be least affected by the upheavals that followed.

Bacteria have existed on Earth for more than 3.5 billion years, and during that time many of them have persisted with ancient ways of life that owe nothing to fresh air or the invigorating warmth of bright sunshine. They can live in extraordinarily testing habitats, such as sulphurous springs and deep-sea vents. Until recently they were thought to reach no deeper than the upper few hundred feet of the Earth's crust. However, after some startling discoveries made when drilling for oil, geologists have found that life exists at depths far greater than this. Bacteria have been found in rock samples collected 1½ miles (2.4 km) beneath the surface, and it now seems likely that they survive down to about 4 miles (6.5 km) beneath land and 6 miles (9.5 km) beneath the seas, at which point scaldingly hot rock limits their further spread. These bacteria live in tiny water-filled pores in rock and

form colonies called subsurface lithoautotrophic microbial ecosystems – SLIMEs for short – which survive by collecting energy and nutrients from the minerals around them. For the bacteria in these SLIMEs the Earth's surface is a remote and alien environment, suffused with toxic oxygen. It is a world as hostile to them as their subterranean world is to us.

Cocooned in solid rock, the deepest SLIME bacteria are completely insulated from the periodic catastrophes that rage high above them and live in a world where change of any kind is immensely slow. More advanced forms of life, living in the sea or on land, are not so lucky. Here the opportunities are greater but conditions more erratic. If disaster strikes, the links between different forms of life ensure that species tumble like a house of cards, with each extinction causing several others.

In the turbulent world following a future mass extinction, the surviving forms of

life would emerge into a period of unprecedented opportunity. Freed from much of the competition they previously faced, they would be able to exploit their environment in new ways; this would create a wealth of species that did not exist before. A tremendous range of biological vacancies would open up; exactly what would fill these vacancies would depend partly on their existing adaptations and partly on luck.

FUTURE EVOLUTION

The results of this renewed struggle for supremacy are impossible to predict, but they include many fascinating possibilities. It could happen, for example, that a future mass extinction might be triggered not by the impact of a meteorite but by the violent outpouring of energy that follows the collapse of a nearby star. This kind of event, called a supernova, would bombard the surface of the Earth with a brief but intense burst of radiation, killing many forms of animal life and also damaging the prospects for leaf-bearing plants. If plants were badly enough hit, the way would then be open for a different kind of life to take their place.

That kind of life might well be a living partnership, perhaps between fungi and microscopic algae. These partnerships already exist in lichens, but in future forms the results could be far more impressive: brilliant green mounds many feet high, perhaps packed close together to form carpet-like forests wherever there was damp ground. Following the supernova, flowering plants would probably still exist, but these fungal forests might be more versatile, and thus gain ground quickly.

The mounds would bear no flowers, but every spring their upper surfaces would turn brown or black as the fungi formed their spores. With the arrival of summer the spores would be released in clouds, blowing away like soot-laden smoke. The empty spore cases left behind could provide a welcome feast for animals – including mound-dwelling birds which,

WAITING IN THE WINGS
Fungi occupy an unobtrusive place on life's stage. In the future a changing world may bring them to the fore.

having no need to fly, would have lost all trace of wings. These birds would normally live within the mounds themselves, and they would certainly share their habitat with many other forms of life, including the few amphibians that had managed to survive the supernova, and also giant descendants of today's earthworms, which would tunnel their way through the soft core of the mounds, ingesting their food as they went.

The world outside this new form of vegetation would probably be one of harsh extremes, with average temperatures colder than today's. After millions of years of supremacy mammals might be a thing of the past – an evolutionary experiment that had run its course. Reptiles might still exist, but they would not be the ones we are familiar with today. These future reptiles could well be warm-blooded, covered with felt-like scales that conserved ...dy heat. Most of them would live on ...t some species might have taken up ...resh water or in the sea. Here their ...ld trap a layer of insulating air, giv-...m a silvery sheen as they dived for ...neath the surface.

...e of these reptiles could look con-...like mammals, perhaps even giving ... live young instead of laying eggs, ...sing their young on milk. But de-...eir size they would not necessarily ...ominant forms of life on land. That ...ight well belong to very different ...f life.

...ution can take new directions after ...xtinction, but the physical laws that affect life will remain unchanged. Among other things these laws mean that insects will never be much bigger than they are today, because the extra power of their muscles would never be able to match the extra weight of much bigger bodies, with their heavy outer cases. But while size has inbuilt restrictions, behaviour does not, and changes in the way insects run their lives could have some far-reaching results.

THE INTELLIGENT INSECTS

In our world many insects depend on flowering plants for their survival. In the future, if flowering plants became much more rare, these insects would have to adapt to new sources of food. Many would continue to live solitary lives, but some – like today's social insects, which include ants, bees, wasps and termites – would function in groups ranging in size from a few dozen individuals to many millions. Among them might be deadly predators, hunting in fast-flying swarms like the aerial equivalent of piranhas and attacking animals many times their own size. They would also include hunters and scavengers, and many other species that used the fungal mounds as nourishment. This would be a world in which insect swarms controlled the skies. Some would skim close to the ground as they searched for food, while others would dive on their fellow insects from above, scattering them like a herd of panic-stricken antelope when lions or cheetahs are suddenly let loose on them.

Long after the disappearance of humans, the nests of these insects would be the biggest and most complex structures made by living things on land. Towering cities, made out of sun-baked clay, would be equipped with landing areas at many different levels above the ground, and perhaps with their own water supplies, much like some termite nests today. At first these cities would be built and run entirely by instinct, but in the far future even this might change. In humans, intelligence is something that belongs to individuals, but in these animals it might evolve as a collective characteristic. Each group would develop its own solutions to the problems of daily life and devise new ways of building, finding food, raising young and making their surroundings an easier environment in which to live.

It sounds like a far-fetched idea, and it certainly has only the slenderest chance of actually coming about. But one thing is certain: if it ever did happen, it would be no more astonishing than our own story, which started so unremarkably several million years ago and eventually transformed the Earth.

INSECT WORLD *Social insects already build the most remarkable structures in the natural world. However, their true potential may still be waiting to be realised.*

INDEX

PICTURE CREDITS

T = top; B = bottom; L = left; R = right

3 Planet Earth Pictures/Ken Lucas. 6 DRK Photo/Larry Ulrich. 7 BIOS/Hubert Klein. 8 Planet Earth Pictures/Flip Schulke, BL. 8-9 DRK Photo/Barbara Cushman Rowell. 10 Bruce Coleman Ltd/Konrad Wothe. 11 Bruce Coleman Ltd/Kevin Rushby, TR; Bruce Coleman Ltd/Harald Lange, BR. 12 DRK Photo/Wayne Lynch, TL; Science Photo Library/Charlotte Raymond, BR. 13 Werner Forman Archive, T; Magnum/Bruno Barbey, B. 14 DRK Photo/Barbara Cushman Rowell. 15 DRK Photo/Larry Ulrich. 16-17 DRK Photo/Richard Longseth. 18 Science Photo Library/John Sanford, T; DRK Photo/M.C. Chamberlain, B. 19 DRK Photo/Tom Bean. 20 Siena Artworks Ltd, London/Pavel Kostal. 21 DRK Photo/Tom Bean. 22 Planet Earth Pictures/Beth Davidstow. 23 Planet Earth Pictures/Krafft. 24 Siena Artworks Ltd, London/Jim Robins. 25 Bruce Coleman Ltd/Mary Plage, T; Planet Earth Pictures/K. & K. Ammann, B. 26 Michael Holford, L; Science Photo Library/John Reader, R. 27 BIOS/ Michel Gunther. 28 DRK Photo/ John Cancalosi. 29 DRK Photo/ Stephen J. Krasemann. 30 Planet Earth Pictures/John Eastcott/ V.V.A. Momatiuk. 31 Auscape International/D. Parer & E. Parer-Cook, TR; Robert Harding Picture Library, B. 32 Robert Harding Picture Library. 33 Magnum/ Abbas. 34-35 AKG Photo, London. 36-37 Magnum/Erich Lessing. 37 Mary Evans Picture Library, BR. 38 AKG Photo, London. 39 DRK Photo/Tom Till. 40 Mary Evans Picture Library. 41 Auscape International/J.M. La-Roque, T; Tom Stack & Associates/Thomas Kitchen, BR. 42 Environmental Images/John Morrison. 43 NHPA/ Anthony Bannister, T; NHPA/ Stephen Dalton, B. 44 *The Electricity Factory of the North at Croix-Wasquehal* by Hippolyte Lety, Musée des Beaux-Arts, Tourcoing/ Bridgeman Art Library, T; Mary Evans Picture Library, B. 45 Robert Harding Picture Library/ Bildagentur Schuster/Ege, TL; DRK Photos/Larry Ulrich, BR. 46-47 Format/Sue Darlow. 47 Magnum/Stuart Franklin. 48 Planet Earth Pictures/Ivor Edmunds, TR; Bruce Coleman Ltd/Geoff Doré, BL. 49 NHPA/ Kevin Schafer. 50 Robert Harding Picture Library/J.E. Stevenson. 51 Siena Artworks Ltd, London/Pavel Kostal. 52 Tom Stack & Associates/ Gary Milburn. 53 Robert Harding Picture Library/Hans-Peter Merten. 54 Sonia Halliday

Photographs. 55 Bruce Coleman Ltd/Hans-Peter Merten. 56 Still Pictures/Mark Edwards. 57 Magnum/Fred Mayer. 58 Environmental Images/Martin Bond, TR; Still Pictures/John Mater, BL. 59 Environmental Images/Martin Bond, TL; Planet Earth Pictures/Bud Smithey, BR. 60 Still Pictures/Edward Parker. 61 Bruce Coleman Ltd/Clive Hicks. 62 DRK Photo/D. Cavagnaro, TL; NHPA/T. Kitchin & V. Hurst, TR. 63 Magnum Photos/Philip J. Griffiths. 64 Magnum/Hiroji Kubota. 65 Mary Evans Picture Library, BL; B. & C. Alexander, TR. 66 Corbis-Bettmann, TL; Topham-Picturepoint, BR. 67 DRK Photo/Tom Bean. 68-69 Auscape International/O. Helmi. 69 Planet Earth Pictures/Adam Jones, TR. 70 BIOS/Jean Roche. 71 Planet Earth Pictures/Ford Kristo. 72-73 Auscape International/Reg Morrison. 73 Tom Stack & Associates/Gary Milburn, BR. 74 Environmental Images/Pete Addis. 75 Magnum/Burt Gunn. 76 Bruce Coleman Ltd/Alain Compost. 77 Environmental Images/Alex Olah, TL; Environmental Images/Dominic Sansoni, TR. 78 Robert Harding Picture Library/Adam Woolfit. 79 Still Pictures/D. Delfino. 80 Tom Stack & Associates/Byron Augustin, TL. 80-81 Tom Stack & Associates/David L. Brown. 81 Magnum/Bruno Barbey, TR. 82 Environmental Images/Robert Brook, B. 82-83 Tom Stack & Associates/Greg Vaughn, T. 83 Environmental Images/Paul Glendell, BR. 84 Robert Harding Picture Library. 85 Environmental Images/Rob Visser, BL; TSA/Tom Stack & Associates, BR. 86 Planet Earth Pictures/Roger de la Harpe, T; DRK Photo/Marty Cordano, B. 87 Bruce Coleman Ltd/Norman Owen Tomalin, T; Science Photo Library/Philippe Plailly, B. 88 *Coalbrookdale by Night* by Phillippe de Louthenbourg, Science Museum/Bridgeman Art Library. 89 *Bedlam Furnace, Madeley Dale, Shropshire, 1803* by Paul Sandby Munn, Private Collection/ Bridgeman Art Library. 90 Magnum/Fred Mayer, TL. 90-91 DRK Photo/Stephen G. Maka. 92 Bruce Coleman Ltd/Gunther Ziesler, T; Still Pictures/David Hoffman, B. 93 Science Photo Library/Philippe Plailly. 94 Still Pictures/Thomas Raupach, BL. 94-95 NHPA/David Woodfall. 95 Planet Earth Pictures/Javier Corripio, TR. 96 Tom Stack & Associates/Inga Spence. 97 Bruce Coleman Ltd/Dr Eckart Pott, TR; DRK Photo/Larry Lipsky, BL. 98 Format/Jacky

Chapman, TL; Science Photo Library/Malcolm Fielding, Johnson Matthey, BR. 99 Science Photo Library/Adam Hart-Davis. 100 The Image Bank/Michael Melford. 101 The Natural History Museum, London. 102 Bruce Coleman Ltd/Joe McDonald. 103 Planet Earth Pictures/John Lythgoe, BL; DRK Photo/Darrell Gulin, TR. 104 The Natural History Museum, London. 105 The Natural History Museum, London, TL; Planet Earth Pictures/John Evans, BR. 106 Bruce Coleman Ltd/Massimo Borchi, TL. 106-7 BIOS/Bruno Pambour. 108-9 NHPA/B. & C. Alexander. 109 Planet Earth Pictures/Robert Hessler. 110-11 NHPA/Paal Hermansen. 112 DRK/Dick Canby, TL; NHPA/B. & C. Alexander, BR. 113 Planet Earth Pictures/Leo Collier. 114-15 Planet Earth Pictures/Jan Tove Johansson. 115 BIOS/Michel Gunther, TR. 116 Tom Stack & Associates/ NASA/GSFC/TSADO. 117 Siena Artworks Ltd, London/Ed Stuart. 118 DRK Photo/Jeremy Woodhouse, TR; BIOS/Frédéric Beauchene, BL. 119 Environmental Images/Dominic Sansoni. 120 Environmental Images/Colin Cumming. 121 BIOS/Gilles Martin. 122 Environmental Images/Pete Addis. 123 NHPA/Christopher Ratier. 124 Bruce Coleman Ltd/Michael Fogden. 125 The Natural History Museum, London. 126 The Natural History Museum, London. 127 Tom Stack & Associates/D. Holden Bailey. 128 NHPA/G.I. Bernard, TL; NHPA/A.N.T., BR. 129 The Natural History Museum, London. 130 NHPA/Jane Gifford, TL; NHPA/Robert Thompson, TR. 131 Seaphot/Planet Earth Pictures/John & Gillian Lythgoe. 132 NHPA/Martin Garwood. 133 The Natural History Museum, London. 134 Tom Stack & Associates/G.S. Kelly. 135 DRK Photo/Randy Trine, T; Science Photo Library/Hans Halberstadt, B. 136 Science Photo Library/ Martin Bond. 137 Science Photo Library/Alex Bartel. 138 Robert Harding Picture Library, BL. 138-9 DRK Photo/M.C. Chamberlain. 139 Science Photo Library/NASA, TR. 140 Roy Williams. 141 Science Photo Library/Martin Bond. 142 Science Photo Library/Restec/Japan, TL. 142-3 Science Photo Library/ Earth Satellite Corporation. 144 Science Photo Library/ Richard T. Nowitz, TR; Science Photo Library/Philippe Plailly, BL. 145 Science Photo Library/ Chalotte Raymond. 146 Science Photo Library/Dr Gopal Murti. 147 Science Photo Library/Oxford

Molecular Biopyhysics Laboratory, T; Science Photo Library/ Lawrence Livermore National Laboratory, B. 148 Bruce Coleman Ltd/Massimo Borchi. 149 Robert Harding Picture Library. 150-1 DRK Photo/Randy Trine. 151 Science Photo Library/Royal Observatory, Edinburgh, TR. 152 Science Photo Library/CNES, 1987 Distribution Spot Image, TL; Science Photo Library/Baker/ Milon, BR. 153 Science Photo Library. 154-5 Bruce Coleman Ltd/Felix Labhardt. 155 Auscape International/Jean-Paul Ferrero.

FRONT COVER: Auscape International/O. Helmi; Collections/John Wender (inset).

77-015-2